shift

A PRACTICAL GUIDE TO ENHANCED SERVICE MANAGEMENT

LEVESQUE | SWAB

Published by Shift Works, LLC

Printed and distributed by IngramSpark.
First Edition
ISBN: [Insert final ISBN]

Printed in the United States of America

Cover and interior design by In the Fade, LLC

ACKNOWLEDGEMENTS

Accomplishing this work would not have been possible without the steadfast guidance, inspiration, and support of the many distinguished leaders, mentors, and colleagues who have enriched our journey over the years. Each of these individuals has been instrumental in refining our understanding of Service Management and the core principles that underpin this methodology.

To the visionary leaders who imparted their insights and set the standard for excellence, we extend our deepest gratitude. Your clear direction and unwavering commitment have consistently motivated us to pursue our goals with purpose and vigor.

To the dedicated mentors who invested time and wisdom in our growth, thank you for nurturing our development. Your patience, insight, and encouragement have not only taught us how to overcome challenges but also how to think critically, adapt creatively, and lead with integrity.

To our colleagues, both past and present, your collaboration, constructive feedback, and shared experiences have profoundly enriched this work. Together, we have navigated challenges, celebrated milestones, and gleaned invaluable lessons that have become the foundation of our methodology.

We extend this acknowledgment to everyone whose unique perspective, thoughtful critique, and unwavering support have contributed to bringing the Shift Service Improvement Methodology to life. It is a testament to the collective wisdom, collaboration, and shared vision of all those who have journeyed with us along this path.

CONTENTS

INTRODUCTION
Why create a service improvement methodology?

The inception of our journey began with a simple yet profound question: How can we develop a practical, streamlined framework for Service Management that is firmly rooted in the most effective industry best practices? Through extensive dialogue and thoughtful consideration, we discerned the need for an approach that is both succinct and all-encompassing. In essence, the creation of a methodology that serves not only as an actionable guide but also as a catalyst for ongoing improvement.

Our professional libraries are filled with reference materials, study guides, and instructional texts meticulously designed to enhance our capabilities. Many of us possess that cherished "go-to" resource we consult repeatedly. In today's era of digital information—where search engines and AI-driven queries deliver knowledge at unprecedented speeds—one might wonder why a new book is necessary. The answer is straightforward: whereas these resources primarily offer solutions, Shift is dedicated to equipping individuals with the critical acumen to ask the right questions and comprehend why those inquiries are essential.

We maintain that the ability to pose incisive questions is as indispensable as possessing definitive answers. When the opportunity to lead is granted to individuals, they invariably aspire to excel, often selected for their proven skills and intrinsic motivation. Shift is designed to nurture this aspiration by offering a refined perspective through which leaders, managers, and team members alike can reassess their roles and contributions. Although Shift does not conform to the conventional mold of a leadership manual, its underlying principles empower readers to operate more intelligently, with an enhanced focus on collaboration, adaptability, and continuous progress.

This methodology is not exclusively for those already in leadership positions—it is intended for everyone engaged in the service environment. Drawing from decades of both individual and collective experiences across a diverse range of industries, Shift distills lessons from formal training, mentorship, and hard-earned experience into actionable insights. By concentrating on the fundamental components of people, processes, and technology, this work offers pragmatic guidance that can be seamlessly integrated into real-world Service Management scenarios.

Our objective is not to supplant existing methodologies but to complement and augment them. Shift builds upon the expertise you already possess, while simultaneously encouraging the relentless pursuit of further knowledge. We passionately believe that the ongoing acquisition of professional certifications and continued education remains vital to the success of service-oriented roles. Whether you are engaging in this work individually or employing it as a discussion tool within your team, our aspiration is that it inspires you to take decisive action, ask the most pertinent questions, and approach challenges with a refreshed perspective.

Ultimately, the impetus for Shift emerged from our own unwavering drive to enhance and innovate within Service Management. This book embodies that journey, and we fervently hope it will incite a similar passion for excellence among its readers. By focusing on formulating the right questions, fostering robust collaboration, and embracing the evolution of both processes and technology, we are confident that Service Management can transcend mere efficiency to become profoundly transformative. It is our sincere hope that Shift will inspire you to continually evaluate, adapt, and improve your strategies, thereby making a measurable and lasting impact on your teams and organizations.

What is Service Management?

Service Management represents an organization's systematic approach to delivering high-quality services that resonate with customer expectations. It embodies a disciplined framework through which entities ensure that every service is meticulously designed, executed, and evaluated. Over time, diverse Service Management principles have been successfully adapted across a multitude of industries—ranging from hospitality to defense/aerospace and countless other sectors. We encourage you to investigate various Service Management methodologies and discern how they might be tailored to your unique area of expertise.

Shift was conceived with the purpose of refining the service lens, zeroing in on critical considerations and core principles that are particularly relevant in today's technology-driven, service-oriented landscape. At its heart, Shift embraces a "whole team" philosophy—one that prioritizes outcomes that foster robust service relationships and drive decision-making. By focusing on tangible results, Shift ensures that teams remain concentrated on delivering benefits that directly enhance customer experiences. The framework is built upon Deliberate Considerations, meticulously designed to align with the specific needs of each mission ecosystem we support, thereby driving services towards their desired outcomes.

Purpose-built to harness the power of data, Shift utilizes rigorous analysis to assess outcomes and continuously refine service processes. Through deliberate dissection, comprehensive analysis, and subsequent reconstruction of processes based on measurable outcomes, Shift empowers teams to leverage their collective skills and propel forward as a cohesive, unified entity. It is only after we have thoroughly evaluated our maximum potential that we can begin to explore the full spectrum of available resources and current infrastructures, striking a balanced synergy among people, processes, and technology. Service organizations adopt Shift to fully capitalize on their inherent capabilities, ensuring that the "whole team" approach consistently exceeds customer expectations in Service Management.

Anatomy of Shift

Shift was conceived as a comprehensive service improvement methodology designed to introduce nuanced modifications—or "shifts"—across the people, process, and technology dimensions within existing service ecosystems. Its fundamental objective is to distill and share the collective wisdom and diverse perspectives of seasoned Service Management practitioners, while offering a robust suite of guidance, strategic considerations, practical tools, and relevant references. This framework is intended to assist organizations in charting a clear roadmap toward delivering enhanced customer-centric services, fostering a deeper understanding of individual roles, team interactions, and the essential processes that underpin exemplary customer service.

The service industry is rife with systemic gaps, often arising from a deficiency in adaptability, a lack of comprehensive knowledge, or constrained resources. In response, Shift endeavors to construct logical, well-defined pathways that managers and their teams can follow to navigate these challenges effectively. Moreover, the influence of people and processes is significantly amplified by the continual advancement of technology—a dynamic that Shift leverages by grounding its methodology in established best practices and industry standards, supplemented by curated recommended readings and resources.

At its core, people constitute the nucleus of many service-based industries, forming the very teams that drive daily operations. The varied insights and experiences accumulated over years serve as the bedrock upon which these teams are built. Recognizing and embracing the singularity of each individual is critical; success is inherently linked to understanding how the interplay between people and processes not only shapes operational efficiency but also fosters personal and professional growth. Over time, the processes we adopt evolve through persistent trial and error, often becoming so interdependent with the human element that they cannot be extricated without compromising overall effectiveness. The introduction of technology further augments this constructive collaboration, exponentially enhancing the capabilities of people and the efficiency, speed, and precision of processes.

By integrating the three core tenets—people, process, and technology—and rigorously refining our Service Management methodology within each domain, we begin to discern the profound interconnectedness of the service ecosystem. A nuanced understanding of these relationships is pivotal in devising Service Management strategies that not only adapt to but also invigorate the existing service environment. Maintaining a holistic mindset is essential for enriching the operational landscape of organizations, their partners, and their customers alike.

Shift articulates the concept of a service ecosystem as an intricate network of service environments that permeate business operations, wherein both customers and organizations inhabit distinct yet continuously evolving contexts. Evaluating these environments through the prism of the Shift core tenets enables organizations to pinpoint strengths and identify growth opportunities across critical dimensions—a concept that serves as a foundational reference throughout this methodology.

The collaborative interplay between the people we work with, the processes that guide our daily operations, and the technology we employ establishes the bedrock of Shift. These elements coalesce to form the ecosystems that drive business success. Individuals bring forth unique perspectives forged through years of lived experience, and understanding the underlying motivators, aspirations, and even the subconscious "iceberg" factors that influence behavior is essential for maximizing individual performance and fostering robust team dynamics. Without such personal insight, Service Managers may struggle to unlock the full potential of their teams.

Similarly, the processes developed and refined over years of iterative practice are deeply intertwined with the human element. In many instances, these processes have become so dependent on the nuanced interactions among team members that neither can operate effectively in isolation. The deliberate creation of new processes—engineered to yield consistent, enduring, and repeatable results—is often precipitated by new requirements or significant alterations to existing protocols.

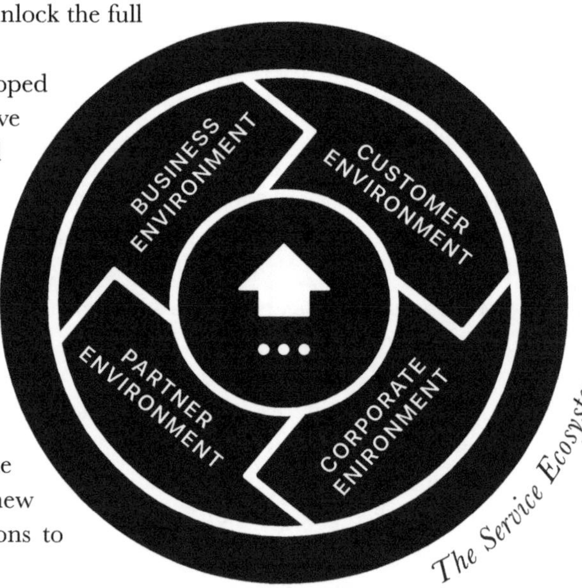

When technology is factored into the equation, its role becomes transformative: it multiplies the inherent skills and attributes of individuals while streamlining processes with enhanced efficiency, speed, and precision. However, the implementation of technology can serve as either a powerful enabler or an impediment, depending on how well it is integrated with human and procedural elements. In this regard, technology completes the foundational triad of

Shift by introducing tangible, material solutions that seamlessly integrate into the broader service environment.

In synthesizing these core principles and continuously honing our Service Management methodology across all three dimensions, we gain critical insights into the profound interconnectedness of our operational environment. This comprehensive understanding is essential for developing Service Management strategies that are both effective and sustainable, creating strategies that not only align with the current ecosystem but also empower it to thrive. Our objective is to cultivate a holistic approach that nurtures the entire service ecosystem, ensuring that every component—from people to processes to technology—is optimized for superior service delivery.

Navigating Shift

The Shift methodology is a comprehensive guide that illuminates the intricate interplay between people, processes, and technology—the three foundational pillars that underpin the success of any organization. By offering a meticulously structured framework, Shift enables leaders and professionals to discern how these essential elements interact to drive service excellence, operational efficiency, and overall business prosperity. Whether you are at the helm of a sprawling multinational enterprise or steering a fledgling startup, the principles expounded in this work provide a universally applicable blueprint for establishing a resilient foundation for success.

The book is systematically organized around three overarching tenets: People, Process, and Technology, each playing a pivotal role in shaping an organization's future. In each section, these core components are deconstructed into clear, actionable principles that can be adapted across various industries to develop practical, robust strategies for implementation. The People section, for example, delves into the transformative power of aligning employee skill sets, values, and personal goals with overarching organizational objectives. It emphasizes the critical importance of stakeholder alignment, skill integration, and purposeful communication as the cornerstones for building cohesive, high-performing teams capable of overcoming complex challenges. Through these insights, readers are encouraged to cultivate a workplace culture that not only fosters individual growth but also propels collective success.

In parallel, the Process section focuses on the optimization of workflows, the enhancement of service delivery, and the establishment of scalable, efficient operations that empower organizations to thrive in competitive markets. Fundamental principles—such as standardization, continuous improvement, and risk management—are elaborated upon to furnish a robust framework that ensures processes remain consistent while also being flexible enough to adapt to rapid changes. By streamlining operations through well-defined processes, organizations can boost overall efficiency and create a fertile environment for innovation and agility. Practical examples and detailed case studies further demonstrate how to construct processes that address both immediate needs and long-term strategic goals, enabling organizations to remain nimble without sacrificing stability.

Complementing these insights, the Technology section delves into the strategic deployment and management of tools and systems that support both people and processes. It provides an in-depth exploration of principles such as technological alignment, system integration, and user-centric design, underscoring the notion that technology should enhance, rather than replace, human effort. When thoughtfully applied, technology serves as a catalyst for efficiency, growth, and innovation. By aligning technological advancements with clearly defined organizational objectives, readers are equipped with the knowledge to leverage tools and systems that drive superior outcomes across all functional areas.

Each section of the book is further enriched by a series of Deliberate Consideration (DC) narratives, which offer actionable insights to guide readers in making well-informed decisions. Contextual tips highlight best practices, delineate common pitfalls, and introduce innovative strategies that can be immediately applied. In addition, Mental Escapes—reflective exercises designed to stimulate creativity, self-awareness, and alignment with organizational goals—provide a unique means of deepening understanding. Comprehensive checklists break down complex tasks into manageable steps, ensuring both thoroughness and consistency, while hands-on exercises afford opportunities to apply the Shift principles in both simulated and real-world scenarios, thereby challenging readers to think critically and innovate.

To fully leverage these features, readers are encouraged to integrate them into their daily routines, engage collaboratively with their teams, and systematically monitor their progress over time. For instance, contextual tips may serve as immediate sources of guidance, while Mental Escapes can spark fresh perspectives and creative problem-solving. Checklists promote accountability and attention to detail, and practical exercises bolster both individual and collective capabilities. Through thoughtful engagement with these tools, readers can unlock the full potential of the Shift methodology, ensuring that their organizations are well-positioned to navigate challenges and seize emerging opportunities.

By focusing on the synergistic relationship among people, processes, and technology, the Shift methodology provides a cohesive, integrated structure that equips leaders and teams to navigate the complexities of today's dynamic environment. Through the deliberate application of its principles and accompanying tools, this comprehensive guide empowers readers to achieve meaningful and sustained improvements in service delivery and management, for an environment of continuous innovation and growth.

Applying Shift Methodology

The service ecosystem is fundamentally defined by the dynamic interplay among three core elements: people, processes, and technology. Each of these pillars wields significant influence, capable of generating both advantageous and detrimental effects. In many cases, these impacts are inherently dualistic, presenting benefits that are accompanied by corresponding drawbacks. For example, the introduction of innovative technology designed to streamline operations might yield an immediate boost in efficiency, yet it can also precipitate unintended disruptions—such as alienating staff who feel overwhelmed or insufficiently prepared for the change.

It is precisely at this juncture that Shift Practitioners must exercise judicious discernment. Recognizing the immediate advantages of a change is merely the initial step; it is essential to evaluate each "Shift" holistically and from multiple perspectives. Often, the positive outcomes are more conspicuous and can easily obscure underlying issues. For instance, a newly implemented process that accelerates service delivery may, in fact, undermine long-term employee engagement if it is not supported by appropriate training or resources.

Consequently, Shift Practitioners are urged to adopt a methodical and systematic approach to every transformation. By engaging in thoughtful planning, attentively observing the subtleties of change, meticulously documenting outcomes, and critically evaluating both the anticipated and unforeseen results, they can avoid being unduly influenced by the allure of initial successes. Seasoned service leaders understand that true, sustainable success is not derived from isolated quick wins but from the continuous cultivation of balanced, positive impacts that resonate throughout the entire ecosystem.

This comprehensive strategy empowers practitioners to respond swiftly to any negative repercussions, addressing issues before they become entrenched and ensuring that the ecosystem remains resilient, adaptable, and aligned with long-term objectives. The goal is to create a harmonious environment in which people, processes, and technology collaborate synergistically to elevate the overall service experience while preemptively mitigating latent challenges.

Moreover, the early identification of potential negative impacts can serve as a critical safeguard, effectively halting further exploration of an option—even when its benefits appear to outweigh its costs. To support this evaluative process, the service ecosystem POWER approach is outlined below, offering a versatile framework that can be applied in various contexts to visualize the key components of change and assess the outcomes during decision-making.

> **PLANNING** a change through to implementation.
> **OBSERVING** the results of the change.
> **WRITING** down or recording the results.
> **EVALUATING** what has been observed and summarized.
> **RESPONDING** by communicating results and best practices.

Utilizing the POWER approach markedly increases the likelihood of success for future change management initiatives by facilitating the creation of a comprehensive register of change plans along with their corresponding observations. This change register is an indispensable instrument in Service Management, serving as a centralized repository that meticulously tracks every modification within an organization's service ecosystem. It offers a clear, unambiguous insight into the nature of the proposed changes, their assigned priorities, and the potential operational impacts they may entail.

By rigorously documenting each change—whether it pertains to processes, technology, or personnel—the register guarantees that no detail is overlooked, thereby enabling stakeholders to monitor progress with precision. This approach not only bolsters transparency but also plays a critical role in risk management. It empowers teams to thoroughly assess both the advantageous and adverse consequences of proposed changes before they are fully implemented, mitigating disruptions and ensuring the continuity of service delivery.

Here is a simple change register template to help you track and manage changes.

CHANGE ID	DATE RAISED	CHANGE DESCRIPTION	PRIORITY LEVEL	STATUS LEVEL	IMPACT	RAISED BY	APPROVED BY	COMMENTS/ NOTES
001	10/12/24	Upgrade to new CRM system	High	In Progress	Positive/ Negative	John Doe	Jane Smith	Implementation in Q4
002	10/05/24	Revise employee onboarding process	Medium	Complete	Negative	Sarah Lee	Mark Adams	Streamlined workflow
003	09/30/24	Integrate new cybersecurity software	Low	Pending	Positive	Paul Brown	Rachelle Green	Awaiting final approval

This change register includes essential columns like Change ID, Date Raised, Change Description, Priority, Status, Impact, Raised By, Approved By, and Comments/Notes to keep track of changes effectively. The table can be modified or expanded, depending on specific needs and circumstances.

THE PEOPLE TENET
ACE in the Hole

Alignment of the whole person, Communication Exchanges, and Engagement with a purpose.

People constitute the quintessential "ace in the hole" within Service Management, underscoring the imperative of holistic personal alignment. When employees are fully aligned both personally and professionally, they are empowered to present their best selves at work—catalyzing innovation, fostering accountability, and enhancing overall satisfaction. The vitality of Service Management is significantly amplified when individuals not only exhibit technical proficiency but also invest emotionally and mentally in the organization's overarching goals. By cultivating environments that prioritize and encourage this complete alignment, leaders are uniquely positioned to unlock the full potential of their teams, ensuring that every member is driven by a shared purpose imbued with passion and dedication.

Equally critical is the role of clear and consistent communication, which serves as the second foundational pillar. Robust communication channels are essential for building trust and fostering collaboration among teams, management, and stakeholders alike. Open and transparent dialogue minimizes misunderstandings and equips teams to address challenges with greater agility and effectiveness.

Furthermore, purposeful engagement acts as the linchpin for sustaining long-term commitment. Employees must perceive that their contributions are both meaningful and intricately connected to broader organizational objectives. As they witness the direct impact of their efforts, their level of engagement deepens, fueling a virtuous cycle of continuous improvement. In the realm of Service Management, such intentional engagement ensures that processes, tools, and strategies are not merely executed passively but are actively and continuously optimized by individuals who are wholeheartedly invested in delivering exceptional service and nurturing a culture of collective success.

PRINCIPLE 1 – ALIGNMENT

Shift places a strong emphasis on the tangible impact that individuals have on service outcomes by highlighting the critical contributions each person makes to the overall service ecosystem. Rather than remaining confined to abstract concepts, this approach rigorously measures and integrates the essential human elements that drive success. At the heart of this methodology is the Alignment principle, which ensures that the skills, values, and goals of all stakeholders are not only fully understood but are also seamlessly woven into the fabric of Service Management strategies. By aligning these core elements, organizations can significantly optimize performance, foster deep and meaningful engagement, and guarantee that every individual works harmoniously toward shared objectives—thereby elevating both the quality and efficiency of service delivery.

Deliberate Considerations (DC)

Deliberate Considerations further illuminate this concept. It is imperative that both the customer's mission and the corporate mission are clearly comprehended, deeply valued, and rigorously respected by all involved stakeholders. Whether articulated in general terms or through detailed descriptions, the mission must always be intrinsically linked to the organization's culture. Authenticity in identifying which values truly matter—those that are instrumental in achieving daily objectives and ensuring sustained success—is essential. Embracing the underlying "why" of the mission not only crystallizes its purpose but also provides steadfast guidance in upholding core values. Additionally, by establishing performance indicators tied to the mission, organizations can align everyday decision-making processes with these fundamental values. Effective leaders remain committed to this course, engaging in deep listening and creating transparent, open environments that allow both employees and customers to connect with and align to the mission's values.

Alignment is about focusing on what truly matters. Service Practitioners are encouraged to care as deeply about the success of their customers as they do about their own achievements. A thorough understanding of individual career ambitions and the cumulative lessons derived from collective past experiences helps to pinpoint the priorities of each stakeholder and team member. It is crucial to discern what is most important to each customer, to identify the people who are significant to them, and to understand the reasons behind these priorities.

• • •

DC 1 - Mission, Vision, and Brand

Alignment with the mission is an indispensable step in comprehending how services influence an environment and are received by its stakeholders. It is imperative that the tasks performed by individuals and the unique skills they bring are visibly integrated into the mission's narrative—ensuring that each person recognizes their role and feels an intrinsic connection to the larger picture. Equally vital is cultivating a sharp vision of where both the people and the mission are headed—envisioning a future that is both aspirational and attainable. This forward-thinking approach demands that the brand be deliberately shaped to reflect the collective identity of the people, the core mission, and the envisioned future, resonating with external audiences as a symbol of success beyond the confines of the current operational environment.

To solidify this connection, it is crucial that branding remains prominently visible and consistently aligned with the intended target. A mission statement, an articulated vision, or a set of guiding principles should be centrally positioned—displayed purposefully in meetings, marketing materials, and customer spaces—to continually reinforce the organizational goal. In service-based enterprises, brand recognition transcends mere notoriety; it embodies a deep understanding of what the brand signifies. Unlike product-based companies, where customers can directly assess the tangible attributes of a product, service-oriented organizations depend on trust, reputation, and nuanced customer perceptions. This reliance renders brand recognition a cornerstone of success, as it shapes how potential clients perceive both the organization and the intrinsic value of its offerings.

When customers encounter your brand, they should immediately grasp its core identity—the exceptional quality of service delivered, the steadfast values that underpin the business, and the promise made to them. A robust brand identity communicates these fundamental elements in a manner that resonates on both emotional and intellectual levels, instilling confidence in customers even before they engage with your services. This familiarity fosters trust, arguably the most crucial element in service industries, where purchasing decisions are often made without the benefit of physically examining a product.

Moreover, strong brand recognition confers a significant competitive advantage. In markets saturated with similar service offerings, differentiation hinges on how effectively your brand communicates its unique strengths. This message could be an unwavering commitment to exceptional customer service, specialized expertise in a niche area, or an uncompromising dedication to ethical practices. Such clarity ensures that customers feel secure in choosing your company over competitors, fully aware of what to expect from your brand.

In the long term, potent brand recognition transforms customers into loyal advocates. When customers forge a deep connection with the essence of your brand, they are more inclined to return for future services and enthusiastically recommend your organization to others. Thus, brand recognition is not only vital for attracting new business but also for sustaining it, thereby laying a robust foundation for enduring success. In service-based companies, employees serve as the living embodiment of the brand—their performance, interactions, and overall service delivery communicate to the world who you are, what you do, and why you are indispensable.

Shift places a premium on developing an in-depth understanding of the customer's mission, vision, and brand. By discerning how the customer defines and prioritizes strategic components, organizations can unlock enhanced opportunities to connect with and exceed expectations. Establishing robust relationships through meticulous research into the internal drivers of the customer's business ensures that services are delivered with consistent precision and insight.

Furthermore, it is essential to maintain vigilant oversight of stakeholder strategies and business objectives, ensuring that the priorities of functional areas remain in continuous alignment with high-level, long-term goals. Sustaining a connection between daily operational activities and the overarching strategic purpose not only perpetuates a winning mindset but also reinforces a collaborative approach where every team member's contribution is valued equally—a philosophy encapsulated by the "whole team" concept.

Effective service delivery begins with aligning individual roles and skills to the broader mission and vision, creating a sense of purpose and direction. Strong branding—rooted in visibility, consistency, and emotional resonance—cultivates trust, differentiates the organization, and drives long-term customer loyalty. Shift emphasizes deep understanding of the customer's mission and strategic goals, ensuring services align with and advance stakeholder priorities through a unified, team-driven approach.

●●●

DC 2 - Skills Inventory

Skills inventories are strategic instruments that empower highly productive organizations to optimize the full spectrum of their workforce's experiences, professional competencies, and qualifications. By systematically cataloging both explicit skills and the subtler, intangible attributes within an organization, these inventories provide a clear, comprehensive snapshot of available talent. This detailed mapping enables leaders to make informed decisions regarding resource allocation, team composition, and the identification of future training needs. Skills inventories transcend surface-level qualifications, offering a holistic view of workforce strengths and pinpointing areas ripe for development.

Typically, skills inventories are constructed through a combination of employee self-assessments, detailed work histories, formal organizational evaluations, and other sophisticated data-gathering methods. Employees are encouraged to reflect critically on both their soft and hard skills, certifications, and accumulated experiences, thereby contributing valuable insights into their individual competencies and areas of expertise. This self-reflective input is then cross-referenced with internal organizational data—such as performance reviews, project outcomes, and peer feedback—to create a well-rounded, nuanced profile of each employee's capabilities.

By methodically mapping these attributes, organizations can more seamlessly align the skills of their workforce with broader business objectives. For instance, a company undergoing digital transformation can leverage its skills inventory to identify employees who possess the technical competencies or leadership qualities required for upcoming projects. This strategic approach not only enhances operational efficiency but also supports robust succession planning by revealing the existing talent pool and highlighting any skill gaps that may need to be addressed through targeted recruitment or tailored professional development programs.

In summary, skills inventories serve as indispensable tools for organizations aiming to maximize their human capital. By maintaining a dynamic, up-to-date understanding of employee capabilities, businesses can ensure they are well-equipped to meet current demands and poised to tackle future challenges, thereby driving sustained organizational success and a lasting competitive edge.

DC 3 - Skills Integration

Integrating diverse skillsets has long been a cornerstone of effective education, where the capacity to interconnect fundamental skills—such as reading, writing, listening, and speaking—not only enriches the learning experience but also enables educators to reach and engage a wide spectrum of learners. This comprehensive, holistic approach to education mirrors a broader organizational truth: aligning individual skills, intrinsic values, and personal goals is essential for achieving a harmonious synergy between a company's mission and the evolving needs of its customers. Much like the way diverse competencies foster a vibrant, enriched classroom environment, a well-curated inventory of skills, values, and objectives plays a critical role in cultivating alignment within an organization.

Values serve as the moral compass that guides both individuals and organizations, fundamentally shaping decisions and actions. While personal values are often deeply ingrained and imbued with significant personal meaning, organizational values may vary—sometimes subtly and at other times markedly. When discrepancies between these sets of values become too pronounced, whether among colleagues or between an individual and the organization, they can lead to dissatisfaction, disengagement, or even conflict. For this reason, it is vital to understand and respect personal values when evaluating and collaborating with individuals or teams. Organizations that dedicate the time and resources to recognizing and honoring these values create environments where employees feel genuinely seen and understood, thereby fostering a profound sense of belonging and loyalty.

Moreover, aligning values raises important considerations regarding how personal beliefs intersect with broader corporate and customer values. An organization that successfully bridges this divide builds a cohesive, value-driven culture where individual convictions harmonize with overarching organizational priorities and customer expectations. This alignment not only optimizes internal operations but also deepens relationships with customers, as the company's core values resonate on a more intimate level with those it serves.

Equally critical are clearly defined goals, which, although distinct from values, are indispensable in achieving this alignment. Whereas values provide the underlying directional force, goals translate these values into actionable steps that both individuals and organizations undertake. Shared goals ensure that teams progress cohesively in the same direction, while individual goals underscore personal development and growth. Whether these goals are organizational or personal, they instill a sense of accomplishment and fulfillment.

Within the Shift methodology, setting realistic yet challenging goals is paramount; overly simplistic goals may sap motivation, whereas excessively difficult objectives can breed frustration. Effective goal setting often necessitates clear and concise job descriptions, enabling realistic, measurable targets to be established within the context of service delivery. When both individuals and organizations pursue meaningful, attainable goals, the resultant sense of progress fuels ongoing engagement and enhanced productivity.

For stakeholders—be they employees, customers, or external partners—the alignment of values and goals is particularly crucial. Their investment in the organization's mission is predicated on a shared understanding of its strategic direction. Although values and goals may adapt to various scenarios, relationships, or objectives, they invariably form the bedrock upon which collective success is built. Without this shared alignment, teams risk losing their clarity of purpose, rendering it difficult to chart a coherent path to success.

By fostering a culture where personal values and organizational goals are in perfect synchrony, companies cultivate environments in which individuals are inspired to contribute, teams operate in unison, and customers are confident in the organization's unwavering commitment to fulfilling its promises. This profound clarity of purpose distinguishes high-performing organizations and is instrumental in securing long-term success.

● ● ●

Routinely survey your teams—not just for the skills they bring, but for how well they feel their personal values and goals intersect with the organization's mission. Use these insights to plan skill-building sessions, adjust job roles, or refine communications. Small, regular check-ins help you spot—and resolve—value misalignments early, ensuring that your team's diverse strengths are fully harnessed and contributing to a culture of shared purpose and ongoing improvement.

MENTAL ESCAPE: INTEGRATING SKILLSETS INTO Service Management

Take a moment to reflect on how integrating essential skills—such as reading, writing, listening, and speaking—enriches your learning experience. Each of these skills plays a vital role in shaping your understanding and facilitating effective communication. Picture yourself immersed in a learning environment where you actively engage in reading materials that challenge your thinking. As you absorb information, you find yourself writing reflections that deepen your comprehension, allowing you to articulate your thoughts clearly.

Now, envision a scenario where listening becomes an active tool for growth. Imagine attending a workshop where speakers share insights relevant to your field. You fully engage, not just passively hearing but actively processing the information, which enhances your ability to connect concepts. Next, picture yourself in a conversation where you articulate your ideas with confidence, demonstrating the power of effective speaking skills in sharing knowledge.

Finally, consider how aligning customer and corporate values with specific goals can amplify your impact. Visualize this alignment as a compass, where values serve as the guiding direction and goals map out the path forward. Think of the tangible benefits that arise when values and goals intersect. This alignment not only enhances corporate culture but also fosters innovation and creativity.

Reflect on individual experiences where you have witnessed this intersection. Perhaps you have been part of a team that rallied around shared values, leading to a successful project that exceeded expectations. Or maybe you have seen firsthand how a commitment to integrity helped build trust with clients, resulting in long-term partnerships. As you continue this Mental Escape, allow yourself to visualize how your own values and goals intersect. Consider how you can leverage your skills—reading, writing, listening, and speaking—to clarify your objectives and align them with your core values. By recognizing the synergy between values and goals, you can embark on a journey toward meaningful achievement that resonates deeply with who you are.

●●●

DC 4 - Management, Leadership & Followership

Management, leadership, and followership represent interdependent yet distinct facets of effectively engaging with people in any organizational setting. The development of these interconnected domains necessitates a deliberate blend of hands-on experience, robust interpersonal relationships, unwavering dedication, and keen self-awareness. Whereas management primarily concerns itself with establishing structure, streamlining processes, and ensuring the meticulous execution of operational tasks, leadership is characterized by the articulation of vision, the capacity to inspire, and the transformative ability to drive meaningful change. It is imperative to acknowledge that excellence in management does not automatically equate to exemplary leadership, and conversely, a leader of exceptional caliber may not inherently possess the skills required for effective management. These roles, while overlapping in certain aspects, demand distinct competencies and nuanced approaches.

Effective leaders must be adept at identifying and harnessing the unique strengths, skills, and capabilities inherent within their teams. This involves not only recognizing individual talents but also strategically aligning these competencies with the organization's overarching goals to optimize efficiency and maximize impact. Successful leadership transcends the mere achievement of objectives; it actively leverages these individual strengths to foster positive, transformative change both within teams and across the organization at large. Moreover, true leadership involves empowering and motivating team members by equipping them with the requisite tools, authority, and confidence to assume leadership roles themselves. This process of empowerment nurtures an environment of collective growth and shared accountability, while sustained motivation is cultivated through consistent encouragement, thoughtful mentorship, and unwavering support, enabling individuals to realize their full potential.

Some of the most distinguished leaders across industries exhibit a remarkable willingness to embrace change and, when circumstances warrant, to be led by others. They understand that leadership is not solely about exerting control but also about recognizing when to listen, follow, and learn. This realization underscores the often-overlooked importance of followership—a critical component of effective leadership. Many emerging managers are taught that to truly excel as leaders, they must also master the art of followership by recognizing that active engagement, humility, and a steadfast commitment to the collective success of the team are essential attributes of both roles.

When examining the qualities that define exceptional leaders and effective followers, several common traits emerge:

- **Courage to Seek Clarification:** Both leaders and followers must possess the fortitude to ask probing questions, even when those inquiries seem elementary, thereby ensuring clarity and minimizing misunderstandings.
- **Acceptance of Diverse Perspectives:** Recognizing that differing viewpoints can stimulate collaboration and drive innovation is crucial; neither leaders nor followers should assume that any single perspective holds all the answers.
- **Constructive Challenge:** The ability to respectfully challenge ideas, propose viable alternatives, and actively contribute to robust discussions is fundamental to enhancing decision-making and problem-solving processes.
- **Leadership Among Followers:** Remarkable followers often exhibit leadership qualities within their peer groups, taking initiative and guiding others without necessarily holding formal authority.

A cornerstone of effective leadership lies in the promotion of consistency within the team. This involves setting a positive example, establishing clear and attainable expectations, providing concise and actionable directives, rigorously following up on commitments, and maintaining open channels for feedback. Leaders who embody these behaviors cultivate a culture defined by reliability and accountability.

The most exemplary leaders recognize that success is inherently a collective endeavor and that the power of teamwork far surpasses individual accomplishment. They eschew the notion of being solitary "lone wolves" in favor of functioning as integral, collaborative members of a cohesive group. By fostering an environment that champions shared ownership of goals and encourages robust collaboration, they inspire their teams to perform at peak levels and achieve success collectively. True leadership, therefore, is a delicate equilibrium of guiding, supporting, and continually learning, all while nurturing a strong, empowered, and unified team.

DC 5 - Development

Professionals are frequently instructed in the fundamental principles of career progression, development, continuous education, and personal growth—cornerstones that are indispensable for individual achievement. However, the Shift methodology reinterprets these concepts by deliberately aligning personal growth with overarching mission objectives, organizational values, and daily operational activities. In this framework, decision-making processes and key performance indicators are designed not only to promote individual advancement but also to drive progress in the overall environment, mission, and the collective welfare of the organization. Growth within a mission-driven context necessitates a service-oriented mindset, where the relentless improvement of products, services, and processes, whether through incremental enhancements or breakthrough innovations, directly bolsters the organization's core mission.

While top-level leaders bear the responsibility for maintaining alignment across the organization, it is equally crucial for smaller divisions, individual offices, and teams to maintain an unwavering focus on these larger mission imperatives. It is all too easy to succumb to tunnel vision when concentrating on specific projects or departmental objectives; yet effective leaders at every level must ensure that everyday activities consistently reflect the organization's overarching mission. This persistent alignment helps forge a cohesive strategy that permeates every layer of the organization, driving success uniformly across the board.

Nevertheless, a critical question arises: what about the people? Often overshadowed by operational targets, career development must be central to any long-term organizational strategy because it is the key to unlocking latent potential within both individuals and the enterprise. Career development comprises a series of deliberate, incremental steps that professionals take to achieve their long-term aspirations. This includes the strategic cultivation of specialized skills through targeted training, robust networking, mentorship programs, and the willingness to embrace new responsibilities. The goal is to achieve a harmonious balance between individual professional growth and the organization's mission. When employees' career aspirations are closely aligned with broader organizational goals, a synergistic effect emerges that benefits both the individual and the collective enterprise.

Career development also presents significant challenges, notably the risk of turnover. As employees mature in their roles, they may eventually outgrow their current positions or seek opportunities elsewhere, thereby potentially destabilizing the organization. This possibility often envokes hesitation among leaders to fully invest in individual growth, driven by fears of disruptive transitions. However, greater peril lies in the failure to invest in people at all, as stagnation can lead to widespread disengagement and a subsequent deficit in innovation.

Robust Service Management principles offer a pathway to mitigate these risks. By fostering a culture of knowledge sharing and actively discouraging the pitfalls of "knowledge hoarding"—where critical insights are confined to individuals—organizations can ensure that valuable information is disseminated widely. Encouraging open communication and cross-team collaboration helps establish a resilient foundation of shared understanding among all employees, thereby minimizing the risk of overreliance on single points of expertise.

In conclusion, the promotion of individual growth, when meticulously aligned with the organization's mission, not only strengthens each employee but also fortifies the organization's capacity to adapt and thrive. By ensuring that career development and personal advancement are inextricably linked with long-term strategic objectives, leaders can cultivate a dynamic, resilient workforce capable of advancing the mission and achieving sustainable success over the long haul. Unchecked tunnel vision—where focus becomes so narrow that broader opportunities or responsibilities are missed—can quietly derail this alignment. That's why it's critical for leaders and teams to regularly engage in exercises like team alignment mapping, which serve as a mental escape from the daily grind, offering a structured way to step back, recalibrate perspective, and ensure each team member's role remains visibly connected to the mission, values, and strategic objectives of the organization.

MENTAL ESCAPE: TUNNEL VISION CHECKUP

Tunnel vision is a phenomenon where an individual becomes overly transfixed on one specific task, goal, or problem to the exclusion of everything else around them. While the ability to focus intensely can be beneficial in many situations, tunnel vision can become a hindrance when it causes people to miss essential information or opportunities. It limits perspective, leading to mistakes, inefficiencies, or missed chances for improvement.

Understanding how tunnel vision manifests can be relatively easy, but avoiding it is often much more challenging. Consider a simple, everyday scenario: You go to the store with a list of items you need to purchase. After picking up a few things, you struggle to find one item. Determined to track it down, you spend too much time wandering the aisles searching for that specific product. In your fixation, you pass by several other items on your list without even noticing them. As a result, you not only waste valuable time, but the store is closing, and you must leave without getting the items you need. This example might seem trivial, but it is a clear illustration of how tunnel vision can cause us to lose sight of the bigger picture—whether in a grocery store or in more complex, professional environments.

In the world of Service Management or any professional setting, tunnel vision can have far greater consequences. When an individual or team becomes overly focused on a single problem or objective, they may overlook other important tasks or opportunities for improvement. This can lead to poor decision-making, project delays, or even missed goals because the broader context is ignored. While maintaining focus is undoubtedly important, it is equally crucial to regularly step back and reevaluate the situation.

So how can you realize when you are falling into the trap of tunnel vision? A few signs include spending excessive time on one task without making considerable progress, neglecting other important responsibilities, or feeling frustrated because the solution to a problem seems elusive. If you notice that you are becoming increasingly fixated on one aspect of your work, it is likely that tunnel vision has taken hold. Another red flag is when colleagues or supervisors point out that you have overlooked other areas of concern—something that may not even register if you're too deep in your focus. Additionally, losing track of time or feeling like you are constantly under pressure to complete tasks last-minute could indicate that tunnel vision is impacting your ability to manage multiple priorities effectively.

So, what can you do about tunnel vision once you recognize it? Here are a few strategies to help avoid working with blinders on:

Regularly Reevaluate Your Priorities: Schedule time to step back and review your overall goals. Are you still on track with your tasks, or has one project consumed too much of your attention? This reassessment will help you maintain a balanced focus and prevent you from fixating on one issue for too long.

Set Time Limits: If you are working on a complex task or problem, give yourself a set amount of time to work on it before moving on to something else. By imposing these time constraints, you prevent yourself from getting stuck in a loop of diminishing returns and can return to the problem later with a fresh perspective.

Seek Feedback from Others: Sometimes, it is hard to notice when tunnel vision has set in, which is why input from colleagues or team members can be invaluable. Regular check-ins or collaborative reviews can help highlight areas you may have missed and ensure you are considering the bigger picture.

Broaden Your Focus: If you find yourself fixated on a specific detail, zoom out and consider how it fits into the larger context of your work. Asking yourself how this task or issue impacts the overall mission or project goals can help shift your focus back to what is truly important.

Take Breaks: Mental fatigue can contribute to tunnel vision. When you are deeply immersed in one issue for an extended period, stepping away for a short break can help clear your mind and allow you to return with a renewed sense of clarity.

Rotate Tasks: Rather than sticking with one task for an extended period, rotate between different responsibilities. This variety not only keeps your mind fresh but also ensures that no single task monopolizes your time or attention.

Tunnel vision is a common challenge in both personal and professional settings, but with the right strategies, it is possible to manage and overcome. By regularly reevaluating your focus, setting limits, and seeking broader perspectives, you can ensure that you are making progress on all fronts and not just stuck in one area. Maintaining this balance will lead to more efficient work, better decision-making, and greater success in reaching your goals.

Exercise: Mapping Team Alignment

This self-assessment exercise helps you evaluate how well your team aligns with your organization's goals. By identifying areas of misalignment, you can focus on actions to improve harmony and efficiency.

Step 1: Reflect on Organizational Goals

Start by reviewing your organization's mission, vision, and strategic objectives. Write down the key goals that resonate with your role or team's responsibilities.

1. What is your organization's mission statement?
2. What are the top three strategic objectives driving your organization this year?
3. How does your team contribute to these objectives?

Step 2: Evaluate Team Alignment

Use these prompts to assess your team's alignment with organizational goals:

1. Understanding the Mission:
    - Does everyone on your team clearly understand the organization's mission and objectives?
    - Are these goals discussed regularly in team meetings or planning sessions?
2. Role Clarity:
    - Are individual roles and responsibilities well-defined and connected to organizational goals?
    - Does everyone understand how their contributions impact the bigger picture?
3. Collaboration:
    - Does your team work cohesively to achieve shared goals?
    - Are there any silos or miscommunications that hinder alignment?
4. Skillset Alignment:
    - Do your current skills align with the needs of your role?
    - Are there areas where additional training or development is needed?

5. Value Alignment:
    - Do your personal values resonate with the organization's mission and culture?
    - Are there any conflicts between your values and organizational practices?
6. Goal Alignment:
    - Are your personal career goals in harmony with the organization's strategic objectives?
    - Do you see clear opportunities for growth within the organization?

Step 3: Identify Misalignments

Reflect on your responses and pinpoint potential misalignments. Use the following questions to guide this step:

1. Gaps and Challenges:
    - Are there areas where your efforts or your team's efforts do not contribute to key organizational goals?
    - What challenges prevent better alignment (e.g., unclear objectives, lack of resources, conflicting priorities)?
2. Opportunities for Improvement:
    - How can you or your team better support the organization's mission and objectives?
    - What changes in communication, processes, or resources might help realign efforts?

Step 4: Create an Action Plan

Develop a plan to address the misalignments identified in Step 3. Focus on actionable and measurable steps:

1. Set specific goals to enhance alignment.
2. Identify resources or training needed to close gaps.
3. Schedule check-ins with team members or managers to review progress.
4. Establish Key Performance Parameters (KPPs) and Key Performance Indicators (KPIs) to measure alignment over time.

Scenarios for Practice

Apply the above steps to the following hypothetical situations to strengthen your understanding:

> **Scenario 1:** Your team is responsible for improving customer satisfaction, but internal feedback shows delays in response times. What steps can you take to align team efforts with the organizational goal of enhancing customer experience?
>
> **Scenario 2:** You have been tasked with a new project, but its objectives feel disconnected from the company's broader mission. How will you approach clarifying and aligning the project with organizational goals?

By the end of this exercise, you will have a clearer understanding of where you or your team stands in relation to organizational objectives and a roadmap for achieving stronger alignment.

• • •

PRINCIPLE 2 – COMMUNICATION

Shift emphasizes the pivotal role that personal interactions play in delivering exceptional service, anchoring its approach in the Communication principle. This principle underscores the necessity of building robust connections through clear and effective exchanges, where trust forms the bedrock of any meaningful dialogue and authenticity ensures that messages are both genuine and resonant. Equally significant is the practical art of crafting messages that are not only clear and purposeful but also directly relevant to the audience. Furthermore, the way a message is delivered—be it through tone, medium, or timing—can substantially influence its overall impact. Collectively, these elements of communication are designed to foster stronger collaboration, enhance service outcomes, and contribute to a more cohesive and high-performing organization.

Deliberate Considerations (DC)

Deliberate Considerations further illuminate the centrality of communication within the Shift methodology. At its core, communication is the process of exchanging information to form ideas, opinions, and thoughts between individuals with the aim of achieving mutual understanding. This exchange is so ubiquitous that people often label communicators simply as either effective or ineffective, a judgment that is sometimes made without appreciating the nuanced complexity of this essential soft skill. In Service Management, where attention frequently centers on tangible outcomes and products, the critical importance of communication is often overlooked. However, fostering robust communication practices can significantly enhance service delivery by ensuring that the subtleties of human interaction are not neglected.

Within the context of Shift, communication is primarily oriented toward people and the achievement of specific goals. A successful approach to communication involves not only conveying information succinctly but also striving for progress toward predetermined objectives. For example, by reducing the number of touchpoints while still ensuring that each engagement is meaningful, organizations can streamline interactions in a manner that builds trust, loyalty, and sustained engagement among stakeholders. In addition, fine-tuning messaging techniques to optimize output can lead to maximized efficiencies, substantial time savings, and potentially exponential growth in service effectiveness, all while utilizing resources judiciously.

Ultimately, Shift's approach to communication—integrating trust, authenticity, clarity, and strategic delivery—creates an environment in which interpersonal exchanges are not merely transactional but are transformative. This holistic view of communication drives enhanced collaboration and paves the way for superior service outcomes across the organization.

• • •

DC 6 - Authenticity

Authenticity, defined as the unwavering commitment to being true to oneself, is a foundational virtue that not only fosters trust but also serves as a cornerstone in defining one's character. At its essence, authenticity embodies an unvarnished honesty, a quality that enables individuals to express their genuine thoughts, emotions, and beliefs without succumbing to external pressures. This intrinsic quality is vital in both personal and professional realms, as it sets the stage for meaningful relationships built on mutual respect and understanding. By being authentic, individuals communicate a sense of reliability and consistency that is indispensable in any environment where trust is paramount.

Beyond the surface-level notion of trustworthiness, authenticity demands a rigorous adherence to one's core values and beliefs, even in the face of adversity. Genuine individuals are distinguished by their steadfast commitment to their principles, choosing not to compromise their integrity merely to conform to societal expectations or to garner favor within a group. This resolute adherence to personal ethics becomes particularly significant in challenging situations, where the temptation to alter one's demeanor to fit in can be overwhelming. In such instances, maintaining authenticity is not only a testament to one's moral fiber but also a powerful statement of individuality that inspires others to act with similar integrity.

Integral to the cultivation of authenticity is the process of deep self-reflection. This introspective journey requires individuals to engage in regular, honest evaluations of their thoughts, actions, and motivations. By developing a heightened awareness of their inner selves, authentic individuals are better equipped to navigate the complexities of interpersonal interactions. They understand that true sincerity involves a willingness to expose one's vulnerabilities, potentially sharing personal experiences and emotions that, while possibly exposing one to criticism or judgment, fosters stronger bonds with others. This openness is essential in creating an environment where trust and mutual respect flourish, as it allows for an exchange of ideas and feelings that are both transparent and deeply personal.

Moreover, authenticity plays a critical role in the dynamics of social interactions, where a delicate balance must be struck between personal expression and social sensitivity. While it is important to allow one's unique personality to shine through, it is equally crucial to remain cognizant of the social context in which one operates. Authenticity, therefore, does not equate to a disregard for social norms or an unfiltered expression of all personal views. Rather, it requires the individual to calibrate their message in a manner that respects the cultural and social frameworks of the environment, thereby ensuring that genuine self-expression does not inadvertently lead to discord. This balanced approach enables individuals to remain true to themselves while simultaneously cultivating harmonious relationships, as it underscores the importance of empathy and contextual awareness in interpersonal communications.

In the broader scope of leadership and organizational culture, authenticity is instrumental in building resilient teams and sustainable business practices. Leaders who exhibit genuine authenticity inspire confidence and loyalty among their teams by modeling behaviors that are consistent with their stated values. This consistency not only enhances the credibility of the leader but also creates a ripple effect throughout the organization, encouraging employees at all levels to embrace their true selves. As a result, an authentic organizational culture fosters a climate of transparency and open communication, where individuals feel empowered to contribute innovative ideas and collaborate effectively. The cultivation of authenticity, through self-reflection, balanced self-expression, and a steadfast commitment to personal values, lays the groundwork for enduring trust, robust relationships, and collective success.

MENTAL ESCAPE: BUILDING TRUSTING RELATIONSHIPS

Building a trusting relationship with those you engage with involves sharing parts of yourself and allowing some vulnerability. It is through this openness that genuine connections are formed. However, this does not mean you need to reveal everything about yourself, it is perfectly fine to keep certain things private. Maintaining personal boundaries is healthy. What matters is that others see you as a real person, not just someone defined by a position or title. By letting people get to know you beyond your role, you foster deeper, more authentic relationships that are deeper than surface-level interactions.

Why do people like you?
Why might people not like you?
What can you do to increase your Authenticity?

To truly maximize the effectiveness of this concept, a Service Manager must develop the ability to read a room and be highly aware of the environment they are in. Understanding the atmosphere and dynamics of a situation is essential to knowing how to engage with others effectively. At times, it might be appropriate to make small talk or discuss common interests to build rapport, but this should always be done with a clear understanding of the audience. A key element of this awareness is recognizing that authenticity is not a one-size-fits-all trait. While it is important for Service Managers to display their personality, they must also be mindful of how much of themselves they share in professional settings.

Authenticity does not mean abandoning professionalism or becoming overly familiar. Service Managers should avoid mistaking someone's time or attention as an invitation to reveal their most "authentic self" without restraint. Instead, they should aim to strike a balance—being genuine and approachable while still maintaining respect for others' boundaries and the context of the interaction. This ensures that they remain authentic without overstepping, allowing them to build trust and rapport without sacrificing professionalism. The goal is to foster connections that are grounded in mutual respect and an understanding of the situation at hand.

•••

DC 7 - Trust

Trust is frequently regarded as a leader's most valuable currency, a vital element that underpins not only individual relationships but also the overall effectiveness of an entire team. In any high-functioning organization, trust is an indispensable condition—where genuine trust exists, effective teams invariably flourish. The art of mastering communication is intrinsically linked to the establishment of trust among all stakeholders; indeed, the symbiotic relationship between trust and communication is a critical cornerstone of Service Management. It is, therefore, essential to cultivate strategies that not only build trust but also foster deep, meaningful connections. Once trust is secured, it serves as evidence of one's honesty, sincerity, and unwavering commitment to never intentionally undermine the team's collective success.

Achieving elevated levels of trust among stakeholders effectively dismantles barriers that often impede productivity and innovation. Trust accelerates decision-making processes, facilitates more effective delegation, and ensures that teams operate at peak efficiency by fully leveraging each member's unique strengths. The development of trust is not a passive occurrence but a deliberate process that is nurtured through consistent social interaction. Key elements—such as clear communication, active participation, steadfast commitment, and demonstrable proficiency—form the bedrock upon which trust is built and maintained.

Dependability, as a corollary to trust, serves as its very foundation. When Service Managers consistently honor their commitments and deliver on promises, they instill a deep sense of confidence in their ability to manage tasks and make judicious decisions. Over time, this unwavering reliability empowers them to guide stakeholders toward novel opportunities, thereby reinforcing trust even further. As trust deepens, relationships are not only strengthened but also evolve into enduring bonds that can withstand even the most formidable challenges. In such an environment, open communication thrives; it accelerates problem-solving, facilitates quicker resolution of issues, and fosters long-term success.

When stakeholders observe this high degree of trust and dependability in action, they are more inclined to reciprocate with openness and transparency in their own communications. This reciprocal openness catalyzes the free flow of information, enabling issues to be identified and addressed expeditiously. Consequently, the self-reinforcing cycle of trust and reliability cultivates a collaborative environment where challenges are more readily surmounted and sustained, long-term success becomes not only possible but inevitable for everyone involved.

Trust is not merely an abstract ideal but a pragmatic, dynamic force that energizes every facet of an organization. Its profound impact on communication, decision-making, and overall organizational cohesion underscores the imperative of nurturing trust as a core organizational asset. By investing in trust-building practices, leaders create resilient, agile teams capable of adapting to change and seizing emerging opportunities, thereby ensuring the continued success and vitality of the organization.

●●●

DC 8 - Practicality

In the Shift methodology, this particular Deliberate Consideration may initially appear to reiterate familiar concepts; however, upon closer examination and application of specific parameters, it reveals a considerably more nuanced and intricate paradigm. Rather than merely engaging in abstract theoretical discourse, Shift insists on prioritizing practical solutions. It challenges Service Managers to rigorously evaluate whether every facet of communication is genuinely practical, ensuring that every comment or recommendation is as actionable and effective as possible. This emphasis on practicality is not merely an academic exercise; it is a critical driver of trust, which in turn underpins all successful communication.

Practical Communication, as defined by Shift, is the deliberate act of transferring information from one individual to another with a clearly established desired outcome. Because the intended effect of the communication is predetermined, Service Managers must concentrate on the feasibility and tangible benefits of the proposed solutions. This focus ensures that ideas are not only well-conceived but also grounded in the reality of available resources and capabilities.

A key component of this approach is the careful assessment and continuous review of the inventory of skillsets among stakeholders. Even the most innovative proposal may falter if there are insufficient human resources or technical competencies to support its implementation. For example, a customer or stakeholder might advocate for advanced technological solutions; however, if the necessary funding or expertise is lacking, such proposals become impractical and potentially counterproductive. Thus, it is imperative to ensure that solutions are not only ideal in theory but also viable within the context of real-world constraints.

Shift defines practicality as a fundamental Deliberate Consideration within the broader spectrum of communication because it serves as the backbone of every project. Practicality has the power to either solidify reputations, enhance client relationships, and improve teamwork dynamics, or to undermine them entirely if overlooked. It demands that Service Managers engage in active listening, articulate their thoughts with precision, communicate expectations clearly, and address conflicts in a professional manner. Meetings, too, should be purpose-driven, with every agenda item contributing to the advancement of the project's goals.

Active listening and clear communication are your greatest assets. Prioritize a disciplined approach to communication by listening intently and expressing your thoughts with clarity. This not only enhances mutual understanding but also builds essential trust required for effective collaborative Service Management.

By steadfastly focusing on practical solutions and aligning communication strategies with tangible outcomes, Shift empowers organizations to transcend theoretical ideals and achieve real, measurable success. This approach not only fosters a culture of trust and reliability but also ensures that every stakeholder is equipped to contribute meaningfully to the collective goals of the organization.

●●●

DC 9 - Delivery

Communication in Service Management must be intentional and strategically designed, with a clear focus on advancing progress and achieving measurable outcomes. A Service Manager bears the critical responsibility of ensuring that communication is not only productive and efficient but also serves as the foundation for building trust across all organizational levels and aligning the team toward a common objective. For communication to yield tangible results, it must be purposefully delivered with the explicit intent of achieving a desired or anticipated outcome. Within this context, two primary approaches stand out: Purpose-Driven Communication and Data-Driven Communication.

Purpose-Driven Communication involves aligning every message with a deeper mission or cause. It transcends the narrow confines of profit-making or superficial recognition by inspiring passion and commitment through an emphasis on a higher purpose. Skilled Service Managers maintain close alignment with their stakeholders' overarching mission, ensuring that all communications serve to propel progress toward that long-term goal. By consistently keeping the broader organizational objective in focus, purpose-driven messaging not only galvanizes teams but also reinforces the fundamental reason behind their collective efforts. For instance, consider an auto manufacturing plant where each station on an assembly line performs a specialized task—whether installing engines, transmissions, or electrical components. Despite the diversity of these tasks, the underlying purpose remains singular: to build a complete, functional automobile that embodies quality and innovation.

In contrast, Data-Driven Communication emphasizes the systematic use of analysis and empirical data throughout the communication process. This approach is particularly effective when stakeholders are oriented toward achieving specific, quantifiable goals. By presenting reliable data, Service Managers can inform, persuade, and engage audiences with robust, evidence-based arguments. Data-driven discussions help to crystallize objectives, align the group around a shared purpose, and drive the team toward purposeful engagement. This method leverages the power of numbers and analytics to create compelling narratives that resonate with logically minded audiences.

The key lies in selecting the communication approach—or the optimal blend of both—that best supports the unique goals of the team, organization, or specific project at hand. These strategies are not mutually exclusive; they can, and often should, be integrated seamlessly to address the multifaceted needs of an organization. By combining purpose-driven and data-driven messaging, Service Managers can create a comprehensive communication strategy that not only informs but also inspires action, driving the entire organization toward sustained success.

●●●

Blend purpose and data for maximum impact.

Striking the perfect balance between inspiring purpose-driven messages and rigorous, evidence-based data can transform your communication strategy. This dual approach not only motivates teams by connecting them to a higher mission but also builds credibility by grounding your message in tangible facts—ensuring your communication resonates on both an emotional and rational level.

In conclusion, effective communication in Service Management is the linchpin of progress. Whether employing purpose-driven messaging to inspire or data-driven techniques to validate and guide decisions, the deliberate integration of these approaches ensures that communication not only conveys information but also propels the organization toward its strategic goals. The Planning Your Communication exercise reinforces this concept through structured practice—blending role-playing, audience feedback, and message refinement—to help individuals master the art of impactful communication. By intentionally balancing purpose and data, Service Managers can foster an environment of trust, clarity, and unified purpose that is essential for long-term organizational success.

Exercise: Planning Your Communication

This exercise is designed to help you practice and refine communication techniques by focusing on purpose-driven and data-driven methods. Through role-playing and feedback, you will gain insight into how to craft and deliver impactful messages in various scenarios.

Objective:

Enhance your ability to communicate clearly and effectively using purpose-driven and data-driven techniques, ensuring alignment with your audience's needs and organizational goals.

Part 1: Preparation

1. Understand the Techniques:
   - Purpose-Driven Communication: Focuses on aligning the message with a deeper mission or cause, inspiring action and engagement by emphasizing the "why."
   - Data-Driven Communication: Relies on facts, statistics, and evidence to support the message, enhancing credibility and clarity.
2. Choose a Scenario: Select one of the following scenarios (or create your own):
   - Scenario 1: Announcing a new team initiative to improve efficiency.
   - Scenario 2: Presenting a progress report to stakeholders.
   - Scenario 3: Addressing a performance issue with a team member.
   - Scenario 4: Pitching a new idea to senior leadership.
3. Prepare Your Message: Write a brief message for your chosen scenario using both communication techniques:
   - Purpose-Driven Version: Emphasize the broader mission or objective.
   - Data-Driven Version: Use metrics, statistics, or facts to substantiate the message.

Part 2: Role-Playing

- 1. Set the Scene:
- Gather participants (or practice with a colleague or mentor).
- Assign roles: speaker, audience, and feedback observer (if possible).

2. Deliver the Message:
   - First, deliver the purpose-driven message.
   - Next, deliver the data-driven message.
3. Audience Role:
   - The audience reacts naturally to each delivery, noting what resonated, what felt unclear, or what could be improved.

Part 3: Feedback and Refinement

1. Feedback Session:
   - The audience and feedback observer provide constructive feedback on:
     - Clarity: Was the message easy to understand?
     - Engagement: Did it capture the audience's attention?
     - Relevance: Did it align with the audience's interests or goals?
     - Delivery: Was the tone, body language, and pace appropriate?
2. Refine Your Approach:
   - Adjust your message and delivery based on feedback.
   - Combine elements of both techniques if appropriate (e.g., start with purpose-driven framing and follow up with data for credibility).

Part 4: Repeat and Reflect

1. Practice Again:
   - Redeliver the refined message.
   - Gather additional feedback to assess improvements.
2. Self-Reflection:
   - What did you learn about the strengths and weaknesses of each technique?
   - How can you apply these insights to future communication tasks?

Evaluation Criteria:
Use the following criteria to evaluate the effectiveness of communication:

1. Purpose-Driven:
   - Did the message inspire action or engagement?
   - Was the broader mission or "why" effectively conveyed?
2. Data-Driven:
   - Was the data relevant and persuasive?
   - Did the evidence support the message clearly?

By the end of this exercise, participants will have practiced and refined their communication skills, gaining confidence in tailoring their messages for different contexts and audiences.

● ● ●

PRINCIPLE 3 – ENGAGEMENT

Engagement lies at the very heart of meaningful interpersonal interactions, serving as the driving force behind robust collaboration and deep connections within dynamic service ecosystems. This profound level of engagement is influenced by several pivotal factors that work synergistically. First, initiative motivates individuals to take proactive steps, fostering a powerful sense of ownership and personal accountability. Delegation, on the other hand, plays a critical role in empowering team members to contribute their unique talents and share responsibility, thereby cultivating an environment where collaborative problem-solving can flourish. Hospitality ensures that every interaction is welcoming and supportive, creating positive experiences that build enduring trust. Finally, availability—defined as being consistently present and readily accessible—enhances relationships by enabling seamless, real-time collaboration. Collectively, these elements not only shape how individuals engage with one another but also significantly elevate service delivery and team performance.

Deliberate Considerations (DC)

Within the Shift framework, the concept of engagement is strategically centered on harnessing communication and alignment to drive results. It involves leveraging a deep understanding of both their teams and customers to enhance the efficacy of every interaction. True engagement extends far beyond simple, transactional exchanges; it requires the deliberate and thoughtful application of interpersonal skills to generate substantive, meaningful outcomes.

This approach emphasizes the practical utilization of people skills to achieve tangible results. Key strategies include effective delegation to maintain project momentum, cultivating an in-depth understanding of team members on a personal level that transcends their formal roles, and taking decisive initiative in response to customer interactions. For instance, when a Service Manager receives feedback indicating customer dissatisfaction, proper engagement would not involve merely acknowledging the issue but also implementing concrete, actionable measures to exceed customer expectations. By consistently going the extra mile, the manager demonstrates that the conversation is far more than a perfunctory exchange—it reflects an authentic commitment to continuous improvement and exceptional customer care. True engagement ensures that every word is substantiated by purposeful actions, driving the organization toward its strategic objectives.

At its core, engagement is the catalyst for building meaningful relationships, driving proactive behavior, and fostering ownership within a team. When individuals consistently show initiative, embrace delegation, offer genuine hospitality, and remain reliably present, the result is more than individual excellence—it is a deeply connected team culture. This foundation of trust and communication naturally transitions into authentic collaboration, where alignment, shared purpose, and mutual accountability enable teams to respond to challenges together. With engagement as the bedrock, collaboration becomes not just possible—but powerful.

● ● ●

DC 10 - Collaboration

An ardent team harnesses the collective resources of its members, working in unison to generate value and solve complex problems while pursuing a shared objective. At its core, collaboration is the dynamic process of uniting diverse talents to achieve one common goal, thereby distinguishing itself from mere cooperation, where individuals assist one another in reaching separate, often unaligned objectives. In true collaboration, every participant contributes to a singular outcome—this is, in essence, the principle of alignment. Within the Shift methodology, collaboration is the critical mechanism by which we engage stakeholders and ensure that every action supports the overarching vision.

Leadership plays an indispensable role in fostering effective collaboration, as key team members must work together to support sound decision-making. It is essential that these leaders not only possess the requisite authority to guide collaborative efforts but also maintain a relentless focus on aligning activities with the organization's mission goals. Effective collaborative teams are built on the foundation of universal buy-in; when all personnel share a sense of value and ownership in the mission, the collective commitment strengthens engagement. This shared commitment becomes especially vital when tasks are delegated and information is disseminated among a larger group, ensuring that every action taken contributes to the common purpose.

● ● ●

Collective buy-in is the catalyst for alignment. When every team member feels a deep sense of ownership, collaboration transcends routine cooperation and becomes a powerful force that streamlines decision-making and propels the organization toward its strategic objectives.

Moreover, successful collaboration demands that all members operate in good faith, characterized by openness, honesty, and sincerity. This means that individuals should readily share both primary and secondary goals, as well as any potential obstacles, while refraining from withholding critical information. The practice of collective ownership over key performance data is paramount, as it guarantees that performance indicators remain valid, attainable, and universally supported by the team. A profound understanding of the organization's mission, coupled with robust collaboration, enables team members to remain consistently aligned during the execution of related tasks.

Furthermore, leaders and other influential team members must cultivate an environment of continuous communication. By maintaining open lines of dialogue, they not only foster a culture of transparency but also encourage regular, meaningful crosstalk among all participants. This ongoing interaction is essential for monitoring the alignment of task performance with mission goals, enabling the team to swiftly adapt to new challenges and opportunities as they arise.

The confluence of effective leadership, genuine collaboration, and continuous, open communication forms the backbone of a resilient team. This integrative approach not only drives superior service outcomes but also creates a work environment where every member is empowered to contribute, innovate, and thrive in harmony with the organization's strategic vision.

MENTAL ESCAPE: A TRUE TEAM

Take a deep breath and imagine a work environment where collaboration drives every action. Picture a space where everyone, regardless of role, feels empowered to contribute ideas and insights. Visualize a team working like a well-coordinated orchestra, with each member bringing their unique strengths and perspectives. Now, feel the sense of collective ownership, where success is shared, and challenges are tackled together.

In this setting, individual roles blur as team members support one another, recognizing that service quality is a collective effort. Everyone feels accountable, and when challenges arise, they are met with a "we," not "me" mindset—no blame, just solutions created through teamwork.

For example, imagine a project deadline looming, and one team member is struggling with a critical task. Instead of letting that person sink under pressure, the team steps in, offering their expertise and assistance to get the task completed. Together, they brainstorm solutions and divide up responsibilities to ensure the project is delivered on time, turning what could have been a failure into a shared success.

This collective approach fosters creativity. Imagine brainstorming sessions where hierarchy fades, ideas flow freely, and collaboration turns suggestions into practical solutions that drive service forward. Clarity and alignment grow from this, with shared goals motivating the team. Each member takes pride in the quality of the service, knowing they have played a part in its success.

When challenges appear, they are seen as shared opportunities for improvement. The team pools expertise, identifies solutions, and strengthens the process together. As this picture develops, feel the satisfaction of working in an environment where teamwork, respect, and continuous improvement form the foundation for success.

• • •

DC 11 - Initiative & Delegation

Initiative is the capacity to assess a situation and independently take action to address it—even when the optimal solution involves delegating responsibility to ensure the task is executed effectively. It is a fundamental quality that can generate a substantial positive impact not only within an organization but also among its stakeholders and customers. While the concept of initiative might appear straightforward, its true potency lies in a deep self-awareness; understanding one's strengths, motivations, and limitations is essential for cultivating and applying initiative effectively.

Initiative Self Check:

- What drives you?
- Do you feel a sense of ownership?
- Do you genuinely care, and why?
- Do you have the necessary resources available?

Within Shift Service Improvement Methodology, initiative and delegation are intertwined as essential mechanisms that foster agility and effectiveness in addressing challenges and capitalizing on opportunities across service ecosystems. Building on the principles of proactive problem-solving and empowered teamwork, initiative is about discerning and addressing needs promptly—without waiting for explicit instructions. Practitioners are encouraged to act autonomously by leveraging their skills and judgment to identify potential improvements or solutions before issues escalate or opportunities vanish. This mindset, often referred to as an "Ownership Mindset," compels individuals to ask, "What can I do right now to enhance the system?" Ownership Mindset not only propels continuous improvement but also integrates seamlessly with the organization's broader strategic goals.

Empowerment plays a critical role in this process, ensuring that team members feel equipped and confident to act on their insights. Leaders are responsible for creating a supportive environment where initiative is recognized and rewarded. In this context, delegation complements initiative by more than merely assigning tasks—it is a strategic alignment of the right individuals with the right resources to maximize impact. Effective delegation requires a careful assessment of each team member's skills, availability, and capacity, thereby ensuring that every task is executed with precision. When tasks are delegated effectively, it demonstrates a deep trust in the team, fosters a culture of accountability and ownership, and significantly enhances overall engagement and morale. Furthermore, adept delegation enables leaders to concentrate on high-level strategic responsibilities while their teams manage operational or specialized tasks, thereby optimizing organizational efficiency.

It is crucial to understand that initiative and delegation are interdependent rather than opposing forces. Successful delegation relies on team members taking initiative with creativity and determination, and environments that effectively delegate tasks empower individuals to assume responsibilities confidently. For example, consider a scenario in which a Service Manager identifies an inefficiency in a reporting system. By exercising initiative, the manager outlines a potential solution and subsequently delegates specific components of the implementation to team members with expertise in data analytics and process design. This ensures that each element of the task is handled with precision, thereby enhancing efficiency and fostering innovation.

Occasionally, it is beneficial to allow an employee additional time to overcome challenges if there is no immediate risk to task completion. Such an approach can facilitate personal growth through adversity. Moreover, a leader's deep understanding of their team is pivotal in assigning tasks effectively. Delegation is a shared responsibility that does not imply merely offloading work without oversight; rather, it involves ensuring that employees understand they are supported and can approach leadership with any questions or obstacles they encounter.

● ● ●

Empowering your team through effective delegation not only enhances operational efficiency but also cultivates a culture of continuous growth and innovation. Trusting team members to take initiative transforms challenges into opportunities for collective success.

By integrating initiative with robust delegation practices, Shift practitioners elevate their Service Management processes in alignment with the methodology's core tenets of adaptability, engagement, and continuous improvement. This integrated approach not only expedites task completion and drives innovation but also lays a solid foundation for scalability and long-term organizational success.

DC 12 - Hospitality

Hospitality is the generous and sincere reception of others, whether they are familiar faces or complete strangers. It involves creating an atmosphere of warmth and trust, ensuring that guests feel appreciated, safe, and valued. When executed with care, hospitality goes beyond mere politeness; it becomes a strategic tool for building relationships, enhancing employee morale, and reinforcing organizational culture. Fostering a sense of belonging is key to any hospitable environment. In any environment, whether a workplace, social setting, or business, hospitality creates an inclusive culture where everyone feels welcome. When employees, clients, or visitors are treated as valued guests, it promotes a sense of belonging. This inclusiveness enhances engagement and encourages loyalty, which can improve employee retention and strengthen team dynamics.

Hospitality and talent development are deeply connected, as both focus on nurturing people. Just as hospitality creates welcoming environments that foster positive experiences, talent development builds supportive cultures that cultivate skills and growth. Both aim to enhance personal potential and satisfaction, driving success for individuals and organizations. When people feel genuinely cared for, they are more likely to invest their energy and creativity into their work. Companies that practice hospitality by offering employee-friendly environments, clear communication, and thoughtful gestures—like recognizing achievements or providing flexibility—are more likely to retain motivated and satisfied team members. Hospitality can set a company apart in a competitive market by making employees feel like valued partners rather than mere resources.

A well-rounded talent management strategy is critical to the long-term success of any organization. This strategy is recognized as a "Talent Triad," comprised of three interconnected pillars: personnel acquisition, development, and retention. Each of these components plays a vital role in not only bringing in skilled employees but also nurturing their growth and ensuring they remain committed to the organization over time. Here is an exploration of each facet of this triad:

Personnel acquisition is the foundation of any successful organization and begins with attracting the right talent. Personnel acquisition is more than filling open positions; it is about strategic recruitment aimed at bringing in individuals who not only possess the right technical skills but also align with the organization's values and long-term vision.

This requires a multifaceted approach:

- Targeted Recruitment: Leveraging data and analytics to pinpoint sources of top-tier talent allows organizations to be selective and efficient. Whether through job boards, professional networks, or direct outreach, focusing on specific skill sets and cultural fit streamlines the hiring process.
- Employer Branding: Today's talent market is increasingly competitive, and candidates often evaluate companies as much as companies evaluate them. A strong employer brand—built on transparency, culture, and opportunities for growth—draws in high-caliber candidates who are motivated by more than just a paycheck.
- Diverse Sourcing: A successful talent acquisition strategy also emphasizes diversity, tapping into a broad pool of candidates from various backgrounds to bring a wealth of perspectives and creativity to the organization.

Talent Development is the second facet. Once talent is acquired, organizations must foster an environment of continuous growth. Development is not merely a benefit—it is a necessity in today's fast-paced industries. Employees want to sharpen their skills and advance in their careers, and organizations must provide the framework to support this. The journey begins with effective onboarding programs that not only introduce new employees to their roles but also integrate them into the company culture. Pairing employees with experienced mentors accelerates their learning curve and fosters a sense of belonging.

Talent Development also focuses on Ongoing Training and Education. In industries with rapidly evolving technologies and methodologies, upskilling is essential. Offering opportunities for continuing education, certifications, and cross-training not only keeps employees competitive but also ensures that the organization remains at the forefront of innovation. Along with training, career pathing creates transparent career progression opportunities that allows employees to visualize their future with the company. Structured paths for growth, whether through promotions, lateral moves, or leadership development programs, keep employees motivated and engaged.

Finally, retention is the third facet of the talent triad. Recruiting and developing talent is only effective if that talent remains within the organization. Retention is the ultimate measure of how well a company supports its employees and meets their needs. High turnover not only leads to direct costs associated with rehiring but also indirectly impacts team morale and productivity. Several key factors contribute to strong retention:

- Recognition and Reward Systems: Regular recognition, whether through performance reviews, bonuses, or simply a company-wide shout-out, demonstrates that the organization values its employees' contributions. Employees who feel appreciated are more likely to stay committed.

- Work-Life Balance: In an era where burnout is common, organizations that promote a healthy work-life balance tend to have higher retention rates. Flexible work arrangements, mental health support, and reasonable workload expectations go a long way in keeping employees satisfied and loyal.
- Employee Engagement and Inclusion: Encouraging employees to take part in decision-making and fostering an inclusive environment where diverse voices are heard contributes to higher job satisfaction. Employees who feel they are part of something bigger are less likely to seek opportunities elsewhere.

While each of these components—acquisition, development, and retention—can stand alone, their true strength lies in their interconnectedness. A company that excels in recruitment but fails to develop and retain its talent will face high turnover and lost potential. Conversely, an organization that excels in development, but neglects acquisition may find itself short of the innovative ideas and fresh perspectives that new hires bring.

By viewing talent management as an integrated structure, organizations can create a continuous loop where top talent is drawn in, developed to its fullest potential, and retained for the long-term benefit of both the employee and the organization. The Talent Triad is a comprehensive strategy that ensures organizations not only attract the best talent but also invest in their growth and well-being, leading to lasting success. An organization that can master this triad builds a competitive advantage by creating a culture of engagement, growth, and loyalty.

Brand reputation is built on how people perceive an organization, and hospitality is a powerful way to influence that perception. Whether interacting with employees, customers, or stakeholders, practicing hospitality ensures that each interaction is positive and memorable. It cultivates trust and leaves a lasting impression, positioning the brand as not only professional but also caring and customer focused. Over time, this leads to a strong, impenetrable brand prestige where people are drawn to the company for both its products and its people-centric culture. Hospitality creates a professional and welcoming environment and can make or break the reputation of an organization. Here are some thoughts on implementing and amplifying hospitality in your environment:

- Addressing by Titles: Acknowledging individuals by their titles or preferred names shows respect and recognition of their position or identity. It demonstrates attentiveness and professionalism, setting a positive tone for interactions.
- Smiling and Eye Contact: Non-verbal cues like smiling and maintaining eye contact convey warmth and openness. These simple gestures make the other person feel heard and valued, reinforcing the hospitality experience.

- Light and Personalized Conversations: Keeping conversations light and focused on the individual helps build rapport. Personalized interactions, such as remembering intricate details about someone's preferences or asking about their well-being, make people feel special and appreciated.

Practical Example in the Workplace

One practical way to apply hospitality in the workplace is to view colleagues as esteemed guests, where their service, well-being, and convenience are always prioritized. Managers can create a hospitable environment by showing empathy, recognizing personal and professional milestones, offering flexible work options, and maintaining open communication. These actions not only boost morale but also create a workplace where employees are eager to contribute, collaborate, and grow.

By cultivating hospitality, businesses can transform everyday interactions into meaningful experiences that foster loyalty, promote inclusivity, and enhance both individual and organizational success.

DC 13 - Production

Within the Shift mindset, production transcends the mere execution of predefined tasks; it represents the culmination of collaborative efforts in which People, Process, and Technology converge to create meaningful, tangible results. This phase is where strategic intent is meticulously transformed into deliverables that directly impact customers and drive operational excellence. At its core, production embodies the very essence of service delivery, it bridges the gap between detailed planning and decisive action, generating measurable, value-driven outcomes that propel an organization forward.

This phase demands a holistic approach, where employees are deeply connected not only to the specific tasks they perform, but also to the broader purpose and objectives of their work. Understanding the "why" behind production is as critical as knowing the "what"; when employees grasp how their roles align with overarching organizational goals and customer expectations, they are empowered to make informed decisions and maintain exceptional standards of quality. Such alignment fosters a profound sense of ownership and accountability, ensuring that the outcomes consistently meet—and often exceed—expectations.

For customers, the production phase is where promises become reality. Transparency and effective communication are pivotal in this dynamic, as they ensure that customers fully comprehend what they are receiving and how it meets their needs. This clarity not only builds trust but also strengthens relationships, enabling customers to recognize the true value being delivered and to gain confidence in the organization's capabilities. Regular feedback loops during production further reinforces this trust, allowing for real-time adjustments that fine-tune outcomes in alignment with customer objectives.

Production in the Shift framework also highlights the critical interplay between People, Process, and Technology. Employees harness their expertise and creativity, supported by robust and adaptable processes, while technology acts as a catalyst for enhanced efficiency and precision. This synergistic integration creates a dynamic environment where innovation thrives, and service delivery continually evolves to meet changing demands. The seamless melding of these elements ensures that production is not simply about completing tasks, it is a powerful driver of innovation and excellence across the entire service ecosystem.

Production is the tangible realization of an organization's vision, where strategic designs take shape to fulfill customer needs and advance organizational goals. In the

Shift mindset, production is more than an endpoint; it is a transformative stage where the collective efforts of the team culminate in outcomes that define success, inspire growth, and pave the way for continuous improvement.

● ● ●

Harnessing the synergy between people, process, and technology is essential.

By integrating diverse expertise with adaptable processes and innovative technology, organizations can ensure that production is not only efficient but also a powerful engine for sustained innovation and excellence, elevating customer experiences and driving long-term success.

DC 14 - Availability

Be present, reliable, dependent, and ready for immediate interaction. Service Management is about service, and making sure personnel and teams are readily available to the customer is key to good service engagement. Stakeholders depend on Service Managers for problem resolution. When team members are unavailable or difficult to contact, then they likely will find alternate means of resolution. A lack of connection between team members could result in a loss of trust or confidence. If not corrected, this type of environment leads to the opposite of what you should strive for, strong engagement, and instead will cause people to disengage and look elsewhere.

Being available can mean a few things. Having an in-person presence in the offices/ locations of your customer can be convenient and offer visible availability. Remote personnel should be "online" and contactable by members of the team and customers. Another important aspect of availability is being available to truly listen and engage with your customer. Customers should feel confident that you hear their concerns and will act. Simply giving head nods/lip service does not instill confidence or trust that you will address their issues.

Presence is about not only being "there" for the customer, but being "there" with the right message, product, or service. It's about making sure a service organization is close enough to their customers to be on their mind in times of need. It is equally important to understand times when they may not be needed and instead be the first person the customer thinks of when they have a requirement or a challenge that they need help with. This approach relates to offering solutions rather than problems, easing their struggles, and aiding their ambitions.

Availability is twice as impactful if you are reliable and dependable.

Reliability and dependability are fundamental pillars of building meaningful and lasting engagement within teams. While these terms are often used interchangeably, there are subtle distinctions between them that highlight their unique contributions to fostering trust and confidence among stakeholders and team members.

Reliability refers to the consistency with which an individual, team, or system performs tasks or delivers results. It is the assurance that something will function as expected, every time, without deviation. Reliability is measurable and often demonstrated through a track record of predictable and high-quality outcomes. For example, a reliable team consistently meets deadlines, produces accurate work, and maintains operational efficiency. Stakeholders value reliability because it creates a sense of predictability, reducing uncertainty in projects and operations.

Dependability, on the other hand, expands on reliability by emphasizing the human and relational aspects of performance. It involves the ability to be counted on in situations that require judgment, adaptability, or extra effort. Dependability includes not just delivering results but also showing a commitment to the team's goals and being responsive to the needs of others. For example, a dependable team member is someone who volunteers to take on extra responsibilities during a crunch period or who communicates effectively when challenges arise, ensuring no one is left in the dark.

While reliability focuses on consistent outcomes, dependability emphasizes the relational and situational nuances of being trustworthy. Reliability can often be quantified. For example, a machine operating 99% of the time or a report delivered on schedule demonstrate tangible metrics. Dependability, however, requires a qualitative evaluation. It is about how a team or individual responds in unpredictable or challenging situations, reflecting resilience, initiative, and accountability.

In practice, the two traits are complementary. A reliable team earns respect through consistent performance, while a dependable team deepens trust by demonstrating loyalty, flexibility, and support during critical moments.

Examples of Dependability for Stakeholders and Team Members

1. Stakeholders:
   - A dependable project manager keeps stakeholders informed about progress, potential risks, and mitigation strategies, ensuring transparency and proactive communication.
   - A dependable vendor not only delivers products on time but also offers solutions when unforeseen supply chain issues arise, showing commitment to the partnership.
2. Team Members:
   - A dependable colleague notices when a teammate is overwhelmed and steps in to share the workload, even if it means adjusting their own priorities.
   - A team leader who fosters an open-door policy and provides guidance during challenges exemplifies dependability, building morale and trust within the group.

Teams perceived as both reliable and dependable create an environment where stakeholders and members feel secure and valued. Reliability ensures that deliverables are met consistently, forming the foundation of confidence, while dependability adds depth by addressing the relational and adaptive needs of dynamic work environments. Together, these qualities reinforce the integrity of the team and sustain long-term engagement.

By focusing on actionable examples and differentiating their impact, leaders can foster both reliability and dependability within their teams, ensuring that trust and confidence remain steadfast in all relationships.

MENTAL ESCAPE: THE ANCHOR OF PRESENCE

Close your eyes and take a deep breath. Imagine standing at the edge of a vast, calm ocean. The horizon stretches endlessly, and the air carries a sense of clarity. The waves are steady, strong, consistent, and dependable—just like your presence in the service environment.

You are the anchor in this ocean. Reliable. Steady. Constant.

As the tide moves in and out, visualize the role you play in your team and with your customers. Your presence is not just about being there, it is about being fully engaged, aware, and ready to act.

Feel the Connection

As you stand firm, imagine those around you—your team, your stakeholders, your customers. They rely on your strength, your guidance, and your ability to be present when challenges arise. They do not need to search for answers elsewhere because they trust you will be there, ready and responsive.

Your availability is more than a status indicator. It's the confidence that when people reach out, they won't be met with silence, delay, or uncertainty. Instead, they receive acknowledgment, solutions, and support. They know that when they need help, you will engage—not just hear, but listen. Not just respond, but act.

The Balance of Presence

Now, shift your focus inward. Picture yourself in a conversation with a customer or team member. You are fully in the moment—not thinking about the next task, not mentally elsewhere, but right here.

You nod with purpose. You reflect before speaking. Your responses are thoughtful, not reactive. Your presence is not just seen—it is felt. This is how trust is built.

As you breathe deeply, feel the weight of that trust settling into your core. Your reliability does not come from being the loudest or the most visible—it comes from your consistency, your follow-through, your commitment to engagement.

Bringing It Back

As you open your eyes, carry this mindset with you. Be the anchor. Be present. Be reliable. Be dependable. When others reach out, be ready—not just to answer but to understand, support, and guide.

Because in Service Management, presence is not just about showing up, it is about showing up with purpose.

• • •

DC 15 - Active Listening

Active listening is frequently heralded as one of the most critical facets of effective communication. It not only facilitates a shared understanding of complex concepts or specific requirements but also plays a pivotal role in nurturing robust relationships among stakeholders and employees alike. Rather than passively waiting for one's turn to speak, active listening involves an intentional, engaged process that requires one to fully concentrate on the speaker's message and its underlying nuances.

At its core, active listening consists of several key components that contribute to its overall effectiveness:

- Listening attentively: Engaging deeply with the speaker's words rather than simply waiting for an opportunity to respond.
- Taking focused notes: Documenting the main points of the conversation to reinforce comprehension and retention.
- Repeating or paraphrasing ideas: Reflecting back what has been communicated either verbally or through written summaries to confirm understanding.
- Asking clarifying questions: Probing further to expand on ideas and eliminate any ambiguities.
- Displaying genuine interest and empathy: Using both verbal affirmations and appropriate body language to show that you value the speaker's contributions.
- Utilizing critical thinking: Analyzing the content thoughtfully to integrate it effectively into your understanding.

In remote environments, however, several challenges can impede active listening. The absence of nonverbal cues, limited opportunities for rich verbal exchanges, and the distractions of multitasking or technical issues can all dilute the quality of communication. Delays in feedback may also be misconstrued as disinterest or inattentiveness, further complicating the interaction. Be aware of these challenges and improvise ways to overcome them.

Active listening is indispensable for ensuring that customer needs are not only understood but also addressed in a way that tailors solutions to exceed expectations. By focusing on the true essence of what customers express, organizations can develop outcomes that foster stronger relationships, enhance collaborative efforts, and drive superior performance. In this way, active listening contributes directly to the delivery of high-quality results, the maintenance of customer satisfaction, and the sustained growth and success of the business.

To cultivate active listening, it is essential to remain fully engaged during every conversation. This means moving beyond simply hearing words to truly absorbing the message being communicated, free from distractions. As an exercise, consider asking yourself reflective questions during discussions: Are you fully present, or does your mind wander? Are you preoccupied with what you plan to say next instead of genuinely processing the information? Such self-awareness is crucial for reinforcing active participation and deepening your connection with the speaker.

One practical strategy to enhance active listening is to maintain consistent eye contact, nod in acknowledgment, and offer thoughtful responses that confirm your engagement with the topic. Active listening transcends verbal communication; it involves interpreting tone, reading body language, and even recognizing the significance of pauses. Even during remote calls without video, consciously using subtle body language cues can significantly improve the quality of your engagement. This attentiveness ensures that the speaker feels both heard and valued, which is fundamental for building trust.

• • •

Reflect on your level of engagement during conversations. Questioning whether you are fully present or distracted can develop greater self-awareness and refine your active listening skills, leading to more meaningful and productive interactions.

Evaluating and continually improving your listening skills not only sharpens your ability to engage with others effectively but also heightens your awareness of when others might not be fully attentive. Over time, this practice of active listening will build trust, boost teamwork productivity, and enhance overall satisfaction within professional relationships, creating a more cohesive and successful work environment.

• • •

THE PROCESS TENET
ACT on the Mission

Adaptability to changing environments, Catalysts that drive process evolution, and Targets that focus on measuring success.

Processes form the backbone of Service Management, providing the essential structure that enables organizations to deliver consistent and reliable outcomes. In today's dynamic and rapidly evolving environments, however, processes must also be inherently adaptable to remain effective. The capacity to evolve in response to emerging challenges, technological innovations, and shifts in customer expectations is indispensable; without such flexibility, rigid processes can quickly transform into bottlenecks that impede progress and stifle growth. Adaptability empowers organizations to adjust swiftly to change, ensuring that their services continue to meet evolving demands and maintain a competitive edge.

Equally important is the recognition and respect for the catalysts that drive process evolution. Whether these catalysts emerge from stakeholder feedback, groundbreaking technological advancements, or variations in market trends, they serve as critical indicators that signal when processes require refinement or overhaul. In tandem with adaptability, organizations must maintain a focused orientation toward the measurable targets that define success—be it key performance indicators (KPIs), customer satisfaction metrics, or operational benchmarks. By aligning processes with clear, quantifiable objectives, Service Management teams can ensure that they are not merely maintaining the status quo but are continually improving and achieving strategic outcomes.

The integration of adaptability and targeted measurement within process management transforms routine operations into a dynamic engine for innovation. This holistic approach guarantees that every element of the service ecosystem is capable of responding to change, thereby fostering an environment where continuous improvement is the norm and where the organization consistently exceeds both internal and external expectations.

Processes are the structural foundation of Service Management, but in fast-changing environments, they must also be flexible to stay effective. Adaptability enables organizations to respond to evolving demands, while attention to catalysts—such as stakeholder feedback or technological shifts—signals when change is needed. When paired with measurable targets like KPIs and customer satisfaction metrics, this dynamic approach transforms processes from static routines into engines of continuous improvement, ensuring services remain relevant, innovative, and aligned with strategic goals.

PRINCIPLE 4 – ADAPTABILITY

The Adaptability principle is fundamental to integrating processes within an environment that is in perpetual flux, ensuring that these processes remain both relevant and effective over time. It centers on the capacity of a team to develop, refine, and adjust its workflows so they can seamlessly align with shifting contexts and emerging demands. By drawing on past experiences, organizations can discern which elements of previous practices are applicable to new or existing workflows, thereby fostering a cycle of continuous improvement. Key aspects of adaptability include the ability to think creatively, make informed and judicious decisions, and design or redesign processes in a manner that is both simple and efficient. These components collectively empower organizations to remain agile and responsive in the face of new challenges and opportunities.

Deliberate Considerations (DC)

In applying this principle, it is crucial to acknowledge that although adaptability and innovation are closely intertwined, they are not synonymous. While they share many traits that contribute to success, each serves a distinct function within an organization. Innovation often acts as a vital subset of adaptability, offering transformative ideas that can be implemented when change is necessary. However, in a service environment that demands continuous delivery, the ability to make swift or incremental adjustments to processes in response to emerging needs or customer demands is generally more effective than halting operations to develop complex, groundbreaking solutions. Adaptability frequently stems

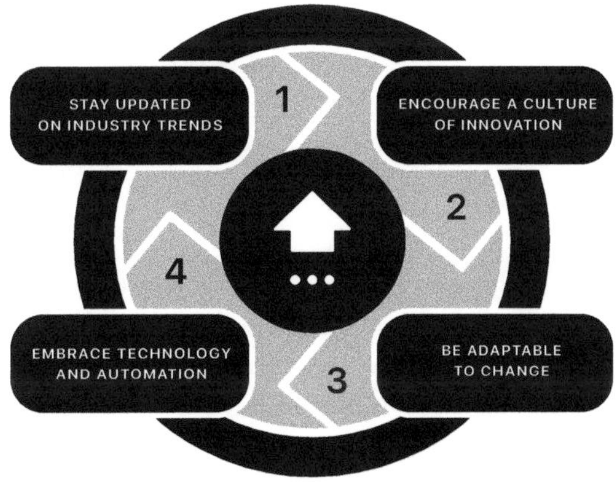

Embracing Innovation and Adaptability

from learned behavior and ensures services continue to operate smoothly while still providing space for innovation.

Moreover, the capacity to adapt both strategies and processes to achieve desired outcomes necessitates a comprehensive consideration of both internal and external influences. A deep understanding of the mindsets of employees, stakeholders, and customers is essential for assessing how effectively a service ecosystem can absorb change and generate success. Overall, adaptability is often distinguished by a keen awareness of prevailing conditions and challenges, as well as an ability to connect these insights to the overarching business objectives of stakeholders. This integration of knowledge not only refines the decision-making process but also cultivates a resilient, forward-thinking organizational culture that is well-equipped to navigate an ever-changing landscape.

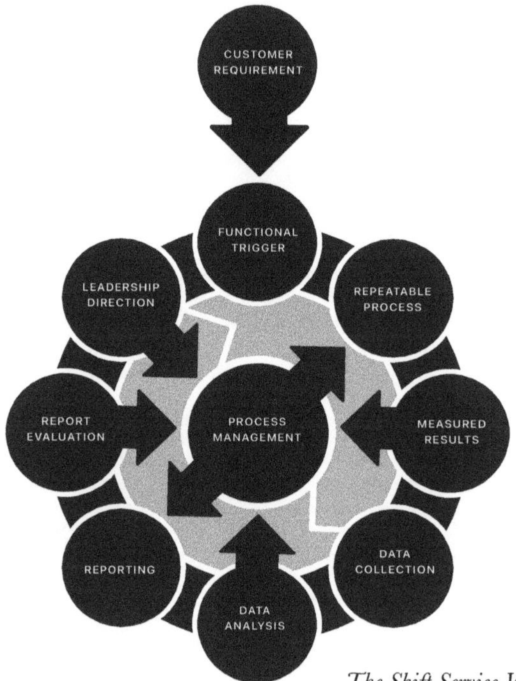

The Shift Service Wheel

The way actions and reactions are molded to the changing service environment defines the level of success in developing processes that easily integrate into the existing service structure (or Service Wheel). A Service Wheel in Service Management represents the continuous improvement and analysis cycle that ensures services remain effective, efficient, and aligned with organizational goals. It is a circular process designed to foster ongoing enhancement through systematic evaluation, feedback, and adjustments. The concept of the Service Wheel emphasizes that Service Management is not a linear process with a definitive endpoint but rather an iterative cycle of refinement.

Key Stages of the Service Wheel:

Assessment: The cycle begins with assessing the current state of services. This involves gathering data on performance metrics, customer satisfaction, and alignment with objectives. The assessment phase helps identify gaps, inefficiencies, or areas where services can be optimized.

Planning: Once the assessment identifies improvement areas, a plan is developed to address these findings. This stage focuses on setting goals for enhancement, allocating resources, and designing strategies for making the necessary adjustments. Key stakeholders and Service Managers align on priorities to ensure the improvements serve the larger organizational purpose.

Implementation: The planned improvements are applied in the service environment, whether introducing new technologies, refining processes, or addressing customer feedback. During this phase, actions are taken to change the service in response to the insights gained from the assessment, and care is taken to communicate the changes effectively.

Monitoring: After implementation, continuous monitoring ensures that the changes deliver the expected benefits. This phase is essential for gathering real-time data on how the improvements are impacting service delivery and whether they are meeting performance goals.

Feedback & Analysis: Establishing a feedback loop is critical. Feedback is collected from customers, stakeholders, and internal teams to evaluate the effectiveness of the changes. This is where analysis of the collected data occurs, providing insights into the success of the adjustments and revealing further opportunities for improvement.

Optimization: Based on the analysis, further refinements are made to the service. This could involve tweaking processes, refining workflows, or introducing additional resources. The optimization phase feeds back into the assessment phase, starting the cycle again.

The Service Wheel is driven by the philosophy of continuous improvement, where each cycle builds on the previous one to enhance service quality. By consistently assessing and refining services, organizations can adapt to evolving customer needs, technological changes, and competitive pressures. Continuous improvement ensures that services never stagnate and are always evolving toward greater efficiency and effectiveness.

Another component is the use of data and analytics at every stage. Decisions are informed by quantifiable results, ensuring that changes are not based on assumptions but on measurable insights. Regularly reviewing performance indicators helps Service Managers make data-driven adjustments, leading to more precise and impactful improvements.

The Service Wheel symbolizes the ongoing nature of Service Management, where processes are consistently analyzed, refined, and optimized in a cycle of perpetual improvement. It highlights the importance of adaptability, continuous feedback, and stakeholder engagement in maintaining high-quality service delivery. This model encourages organizations to remain proactive in their approach, ensuring that Service Management evolves alongside changing demands and expectations. The ability to deliberately change an approach and related processes according to conditions and challenges is fundamental.

The Shift Service Wheel begins with a customer requirement or leadership directive that initiates a functional trigger, setting the process into motion. This triggers a predetermined work cycle within a specific functional area, designed to follow a repeatable process that ensures consistent outputs. Throughout this cycle, measurements are generated to assess process performance and the quality of the resulting product or service. These measurements are stored in a data collection repository, creating a historical record for data mining and further analysis. The collected data is then compiled and filtered to meet specific calculation and metric requirements, enabling a clear understanding of performance trends. Metrics are subsequently reported and evaluated against established goals and performance criteria, with trend analysis highlighting areas that may require corrective actions. Based on these insights, leadership reviews the findings and implements necessary adjustments to enhance performance, ensuring continuous improvement and alignment with organizational objectives.

●●●

DC 16 - Optimization

Optimization is the deliberate process of enhancing efficiency and effectiveness to make the best use of resources, time, and effort in achieving organizational goals. Within the Shift Service Improvement Methodology, optimization goes beyond incremental improvements; it is a comprehensive approach to aligning people, processes, and technology to create value, reduce waste, and drive meaningful outcomes.

At its core, optimization in the Shift context requires a continuous evaluation of how resources are deployed to achieve desired results. This includes examining workflows, identifying redundancies, leveraging technology, and ensuring that human capital is utilized to its fullest potential. The goal is to create a service ecosystem that is not only responsive but also scalable, efficient, and capable of exceeding customer expectations.

In Shift, optimization should be a guiding principle in every decision-making process. This mindset challenges teams to ask critical questions at every stage of a task or project:

- Are resources utilized to their full potential?
- Can this process be simplified without losing quality or effectiveness?
- Is there a way to achieve this goal faster or with fewer inputs?
- How can technology amplify the effectiveness of the process?

These questions encourage proactive thinking, pushing teams to consider innovative solutions and systemic improvements rather than settling for the status quo. Seeking improvements and better processes becomes a cultural norm, enabling teams to think critically about their actions and their impact on the broader service ecosystem.

Optimization manifests across various areas of business, each contributing to enhanced efficiency and effectiveness. Resource allocation involves prioritizing resources based on their impact and necessity, ensuring maximum utility and alignment with organizational goals. Process streamlining focuses on removing bottlenecks and redundancies in workflows, enabling faster execution and greater accuracy. Through technological integration, businesses can automate repetitive tasks or implement advanced tools that amplify productivity and free up human capital for higher-value activities. Cost efficiency emphasizes identifying and reducing unnecessary expenses while maintaining or improving the quality of outputs. Finally, scalability ensures that processes are designed to accommodate growth without requiring significant additional resources, allowing organizations to expand seamlessly and sustainably. Together, these optimization efforts create a foundation for improved performance, customer satisfaction, and long-term success.

These efforts aim to reduce costs, increase profits, conserve resources, and maximize the production of goods or delivery of services. In a Shift ecosystem, the focus is not just on efficiency but also on creating sustainable and adaptable solutions that align with long-term organizational goals.

Adaptability and optimization are intrinsically linked within the Shift framework. While adaptability focuses on adjusting actions and processes to meet the demands of a changing environment, optimization ensures that these adjustments are executed in the most effective and efficient way. In essence:

- Adaptability is about how well teams respond to change.
- Optimization is about how effectively they implement those changes to maximize value.

As customer requirements evolve and external factors such as market conditions or technological advancements shift, businesses must adapt quickly to maintain relevance. Optimization ensures that these adaptations are not reactionary but are instead deliberate, resource-conscious, and aligned with strategic objectives. For example:

- A service team adapting to new customer requirements might implement a streamlined communication tool to replace email chains. Optimization ensures that the tool is effectively integrated, reducing miscommunication and improving response times.
- A production line adjusting to supply chain disruptions might adopt alternative materials. Optimization ensures that the transition minimizes waste and maintains product quality.

Optimization in the Shift context requires a deliberate focus on several key principles. Data-driven decision-making involves using metrics and analytics to identify inefficiencies and measure improvements, ensuring that changes are based on quantifiable insights. Continuous improvement emphasizes the regular evaluation of processes and outcomes to uncover opportunities for refinement and enhancement. Collaboration and feedback are essential, as engaging stakeholders at all levels fosters alignment and encourages the discovery of innovative solutions. Technology acts as an enabler, providing tools to automate tasks, enhance precision, and support scalability, allowing organizations to achieve more with less effort. Sustainability ensures that optimization efforts align with long-term goals, avoiding resource depletion and minimizing the risk of creating new inefficiencies. Together, these principles create a framework for effective and lasting optimization within the Shift methodology.

By integrating optimization into every aspect of service delivery, organizations can ensure that they remain competitive, agile, and effective in meeting both current and future demands. In the Shift ecosystem, optimization is not a one-time effort but an ongoing commitment to excellence, adaptability, and value creation.

●●●

Optimization is not about stretching resources—it is the discipline of improving what truly matters. When people, processes, and technology are purposefully aligned, efficiency becomes inevitable.

DC 17 - Problem Solving

The ability to adapt is a cornerstone of effective Service Management, enabling teams to respond to ever-changing demands, unforeseen challenges, and evolving customer needs. However, adaptability alone is not enough; it must be paired with strategic problem-solving to ensure that adjustments lead to meaningful and sustainable outcomes. While adaptability focuses on recognizing when change is necessary and adjusting processes to meet new conditions, problem-solving dives deeper into identifying root causes, analyzing potential impacts, and crafting targeted solutions. The transition from adaptability to problem-solving represents a progression from reactive adjustments to proactive strategies, where challenges are transformed into opportunities for growth and innovation. By integrating the flexibility of adaptability with the deliberate methods of problem-solving, teams can navigate complexities with confidence and precision, ensuring resilience and continuous improvement within the service ecosystem.

Anyone can solve a problem.

Shift emphasizes not only addressing problems but also thoroughly analyzing their many facets before making recommendations. While problems are often observable, they are not always perceived as actionable. Presenting a problem without the intent to devise or assist with a solution should only occur when limited knowledge of the problem area would hinder productive participation. In most cases, the best approach involves addressing the problem directly by gaining a full understanding of its nature, assessing its impact, identifying key stakeholders for involvement, and developing a clear path to success with stakeholder buy-in. Once a comprehensive problem statement is established and a capable team is assembled, a variety of methodologies can be employed to craft potential solutions.

A well-calculated solution must balance the intended benefits with potential drawbacks for the customer. If implementing the solution requires additional resources—such as cost, personnel, or materials—it is critical to ensure that the outcome is clearly beneficial and well understood by all stakeholders. When required components are unavailable, the solution's scope should be renegotiated to make efficient use of existing resources. This approach ensures that solutions are both practical and impactful, aligning with the Shift principle of continuous improvement and fostering resilience within the service ecosystem.

A Root Cause Analysis (RCA) is a systematic process designed to uncover the underlying causes of problems, ensuring they are addressed to prevent recurrence. Various methodologies guide teams through this process, each offering unique perspectives for identifying root causes and implementing effective solutions. One commonly used technique is the 5 Whys, which involves repeatedly asking "why" until the core issue is revealed. This straightforward approach is particularly effective for problems stemming from process flaws or human error. Another method is the Fishbone Diagram, or Ishikawa Diagram, which visually maps potential causes by categorizing them into major contributing factors such as people, methods, machinery, or materials. This structured exploration enables teams to examine multiple aspects of a problem systematically.

Failure Mode and Effects Analysis (FMEA) provides a structured framework for identifying potential failure modes within a process or system and evaluating their severity, occurrence, and detection. This allows teams to prioritize and address the most critical issues. Fault Tree Analysis (FTA), a top-down deductive approach, breaks down system failures into plausible causes using a tree-like structure. This method is particularly useful for analyzing complex systems with interdependencies. Pareto Analysis, based on the 80/20 rule, focuses on identifying the small number of root causes responsible for most of a problem, enabling teams to address the most impactful factors efficiently. Each of these methodologies provides a distinct lens for problem-solving, and selecting the right one depends on the complexity of the issue and the environment.

Complementing RCA, Cost-Benefit Analysis (CBA) is a decision-making tool used to evaluate the financial and non-financial impacts of a project or decision. By systematically comparing costs and benefits, organizations can determine whether an initiative is worth pursuing. CBA involves several steps: identifying and listing relevant costs and benefits, quantifying them in monetary terms where possible, assigning a period to understand the short- and long-term impacts, and calculating net benefits to assess viability. Additional considerations, such as non-monetary factors like social or environmental impacts, ensure a well-rounded analysis. Sensitivity analysis tests the robustness of results under different scenarios, offering deeper insights into potential outcomes.

The advantages of CBA include providing a clear financial picture, helping prioritize projects, balancing short-term expenses with long-term gains, and highlighting risks and trade-offs. These insights enable more efficient and effective resource allocation decisions, ensuring investments deliver maximum value.

Presenting these findings in a structured decision-making framework is critical for guiding stakeholders. A well-crafted decision presentation begins with a clear objective, provides context through relevant data and challenges, and outlines available options with their respective pros and cons. Analysis, supported by data and tools such as risk matrices or a SWOT analysis, strengthens the credibility of recommendations. Transparency about risks and trade-offs builds trust and demonstrates strategic thinking. Inviting feedback fosters collaboration, often revealing additional insights or considerations. Visual aids and storytelling techniques enhance engagement and make complex information accessible.

Concluding the presentation with a clear outline of next steps, timelines, and responsibilities ensures alignment and momentum toward implementation. By integrating detailed analysis with clear communication, an effective decision presentation empowers stakeholders to make informed choices, driving organizational success and fostering confidence in the process.

●●●

DC 18 - Simplicity

Transitioning from adaptability, it is essential to consider how simplicity plays a pivotal role in process management. The famous American cartoonist, Rube Goldberg, is renowned for designing intricate chain-reaction machines that quite humorously incorporate a multitude of superfluous elements that could introduce significant operational issues over time, particularly with prolonged use.

Simplistic processes, in contrast, offer clarity and ease of repetition while providing a solid foundational framework upon which further improvements can be built. The steps used to enhance these foundational processes should be purposefully designed to achieve equilibrium within the working environment. An excess of subprocesses, outputs, documents, and ancillary tasks can overwhelm a system, rendering it convoluted and inefficient.

Conceive a process with intelligent design, ensuring it aligns with the objectives of both new and existing services while preserving the delicate balance required within a service environment. Maintaining or enhancing the ecosystem is vital to achieving customer goals and objectives, as challenges inevitably arise that may compromise the efficiency, effectiveness, and overall functionality of services provided to stakeholders. It is incumbent upon Service Managers to sustain this balance while ensuring that all communications remain pragmatic and easily understood.

The capacity to distill complex challenges into their most fundamental elements necessitates a straightforward approach. Employing simplistic techniques does not imply ignoring complexity; rather, it advocates for the use of clear and concise Tools, Tactics, and Procedures (TTPs) to aid team members in comprehending, evaluating, and resolving intricate problems. By leveraging complementary aspects of the Adaptability Principle, Service Managers can craft collaborative solutions and develop processes that successfully achieve key objectives.

To effectively accomplish this, evaluate processes in a deliberate and holistic manner. The outcomes of provided services should be consistently assessed through measurable metrics observed throughout the entire process lifecycle. TTPs that delineate the ideal approach to a process improvement cycle—from initial observation to final resolution—are far more effective when they are widely socialized and readily accessible to all employees.

• • •

Clarity through simplicity is a powerful catalyst for success. Establishing and documenting straightforward, repeatable processes not only reduces uncertainty and frustration but also fosters an environment where continuous improvement becomes a natural part of the organizational culture.

Improving a process relies fundamentally on open communication and collaboration at all levels. Employees must recognize that they hold an integral and equal role in the creation and refinement of processes. Adhering to established procedures for creating and documenting new processes can significantly mitigate uncertainty by providing a familiar approach right from the onset. Although these procedures are designed with inherent flexibility and may occasionally deviate, their primary purpose is to seamlessly link the initiation of a process with its eventual completion.

Whether one is developing a new process or evaluating an existing one, it is advisable to implement a repeatable method for process mapping, particularly for tasks that recur frequently. Begin by clearly defining the process to be mapped, then identify its start and end points. Collect pertinent data about the current process through interviews, observations, and existing documentation. Once assembling the data, break the process down into major steps or stages, listing each sequentially and determining the most appropriate format for your map. Common tools for this purpose include flowcharts, swimlane diagrams, or process flow diagrams, with software options such as Lucidchart, Visio, or even basic drawing tools proving useful. Consistently use standardized symbols—ovals for start/end, rectangles for steps, and diamonds for decision points—and share the completed map with stakeholders or team members for feedback and necessary adjustments. Upon finalizing the map, scrutinize it for bottlenecks, redundancies, or inefficiencies, and identify potential improvement opportunities. Ensure that the process map is documented, easily accessible, and communicated to all relevant parties for training or reference.

The final, indispensable aspect of managing processes is Continuous Improvement. This involves regularly reviewing both the process and its corresponding map for updates. The Shift methodology emphasizes integrating these evaluations across the Service Management environment, acknowledging that processes often exhibit observable inefficiencies and are subject to evolution over time.

• • •

DC 19 - Automation

Automation plays a pivotal role in modern Service Management, serving as an indispensable tool for enhancing operational efficiency, reducing costs, and liberating personnel to focus on strategic, value-driven initiatives. In today's fast-paced environment, integrating automation into core processes enables organizations to streamline operations, minimize repetitive tasks, and boost overall productivity. Its true value lies not merely in saving time but also in freeing up human resources to engage in higher-order functions that require critical thinking, creativity, and informed decision-making.

At its essence, automation involves applying advanced technology to perform tasks and functions that were traditionally executed manually—such as data processing, report generation, or routine system checks. These automated processes yield consistent outputs with minimal human intervention, thereby reducing the likelihood of errors and ensuring reliable results. However, before effectively deploying automation, there must first be a well-defined and meticulously documented process. This foundational step is critical, as it allows organizations to identify precisely where automation can deliver the most value. Without a clear understanding of the workflow, automating processes could introduce inefficiencies or overlook crucial steps, resulting in adverse outcomes.

Once clearly defining a process, automation can be applied strategically to conserve both time and resources. For example, automating tasks such as manual data entry, routine system updates, and repetitive reporting not only alleviates the burden on employees but also enhances overall operational efficiency. In Service Management, this strategic shift enables staff to focus on problem-solving, innovation, and customer engagement rather than being encumbered by mundane administrative duties.

Another significant advantage of automation is its capacity to ensure that essential information is readily accessible to all necessary stakeholders. Automated data exports, for instance, guarantee swift and accurate dissemination of critical information, thereby eliminating bottlenecks, and expediting decision-making. Additionally, by reducing the time spent on generating reports, automation ensures that vital data is available exactly when needed, further enhancing operational agility.

Beyond its immediate benefits of time savings and cost reduction, automation also adds value over the long term. As processes become more streamlined, organizations can allocate resources more efficiently, allowing for greater scalability and responsiveness to evolving customer needs. Moreover, automation ensures consistency in output, thereby maintaining exact standards of quality in service delivery and reinforcing customer trust.

●●●

A well-documented process is the linchpin of effective automation. Ensuring that your processes are clearly defined and meticulously documented not only prevents inefficiencies but also enables you to pinpoint the most optimal areas for automation. This critical foundation allows for the sustainable improvement of productivity and quality across your organization.

In conclusion, automation is an essential component of effective Service Management. When employed thoughtfully and strategically, it reduces costs, liberates valuable human resources, and increases overall organizational efficiency—while ensuring consistent and accurate performance of core functions. However, the successful implementation of automation hinges on the prior establishment of a clearly defined process that can fully leverage its benefits while mitigating potential pitfalls. As technology continues to evolve, automation will increasingly become a driving force in the future of service delivery, enabling organizations to maintain a competitive edge and achieve long-term success.

• • •

DC 20 - Best Practices & Methods

Best practices and methods provide a dependable foundation for success, serving as a critical launching point for addressing challenges and implementing new processes. By leveraging approaches that have been proven effective in the past, organizations can significantly reduce the time and resources required to develop solutions from scratch. While many of these established methods continue to be valuable over time, they often necessitate adjustments to stay aligned with current industry trends, technological advancements, and specific organizational contexts. Moreover, exploring best practices from a wide array of industries—rather than confining perspective to a single field—can reveal innovative insights and uncover strategies that might otherwise remain undiscovered. This cross-industry approach underscores the interconnectedness of Service Management, acting as the essential bridge that links customers, companies, and related sectors.

In instances where no relevant best practices exist, organizations have the opportunity to devise new methods tailored to their unique processes or projects. Such bespoke approaches should be meticulously documented to ensure clarity and consistency. Detailed documentation serves a dual purpose: first, it aids team members in replicating and refining the process, and second, it allows the organization to disseminate successful methods across departments or projects, thereby maximizing their overall value. By making these tailored approaches readily accessible, companies foster a culture of collaboration and continuous improvement. This not only enhances efficiency but also ensures that the knowledge gleaned from unique experiences contributes to broader organizational success.

• • •

Embrace a cross-industry perspective.

Drawing on best practices from diverse industries can spark innovative ideas and strategies, helping your organization break free from conventional thinking and achieve transformative success in Service Management.

In summary, best practices and methods are not static templates but dynamic tools that, when adapted and documented effectively, can drive continuous organizational improvement and create a resilient framework for addressing both present and future challenges.

MENTAL ESCAPE: EXPLORING BEST PRACTICES

Take a moment to step back from your day-to-day tasks and focus on two best practices that can enhance your team's effectiveness and collaboration.

First, imagine conducting regular project retrospectives. Picture yourself and your team coming together after each project to openly discuss what went well, what did not, and how to improve for the future. In this space, everyone has a voice, and feedback flows constructively, allowing for continuous learning and growth. Visualize how these retrospectives help identify small adjustments that lead to big improvements in team performance and project outcomes. Next, consider the impact of implementing a robust system for knowledge sharing. Envision a well-organized, accessible platform where team members can document key learnings, processes, and best practices. Now, think about how this system fosters collaboration, allowing team members to easily access valuable information, share expertise, and avoid redundant efforts. This sharing of knowledge strengthens the team's collective capability and ensures that expertise is never siloed.

As you reflect on these two practices, feel how they create an environment of continuous improvement, open communication, and collective learning, driving your team toward greater success and efficiency. Project Retrospectives apply to various contexts beyond project management, such as in personal development or team meetings. For instance, regular self-reflection sessions can help individuals identify personal growth areas and track progress. Similarly, incorporating retrospective principles into departmental reviews can improve overall organizational effectiveness. Knowledge Sharing Systems and practices can be extended to other organizational functions, such as customer service or inter-departmental collaboration. Creating forums or platforms for knowledge exchange can enhance problem-solving and innovation across the company. In customer service, sharing insights about common issues or effective solutions can improve service quality and efficiency.

Exercise: Adaptability in Action – Simulating a Process Adjustment

This exercise allows readers to practice adaptability by applying the Service Wheel concept to address a hypothetical business challenge. Through a live or tabletop simulation, participants will identify gaps, adjust processes, and evaluate the impact of their changes in real-time.

Objective:
Develop adaptability skills by using the Service Wheel framework to evaluate and adjust processes in response to evolving business needs.

Part 1: Setup

1. Understand the Service Wheel: Review the key stages of the Service Wheel:
   - Assessment: Identify gaps or inefficiencies.
   - Planning: Design improvements.
   - Implementation: Apply the changes.
   - Monitoring: Evaluate the impact.
   - Feedback & Analysis: Gather insights for further refinement.
   - Optimization: Make final adjustments to enhance effectiveness.

2. Scenario Selection: Choose or create a hypothetical challenge. Here are some examples:
   - Scenario 1: A new software update disrupts your team's workflow.
   - Scenario 2: A sudden increase in customer inquiries overwhelms the support team.
   - Scenario 3: A key supplier fails to deliver materials on time.
   - Scenario 4: A regulatory change requires immediate process adjustments.

3. Team Assignment: If working in a group, assign roles:
   - Process Owner: Oversees the current workflow.
   - Stakeholder: Represents impacted employees or customers.
   - Analyst: Tracks data and metrics related to process performance.

- Facilitator: Guides the discussion and ensures progress through the Service Wheel.

Part 2: Simulation

1. Assessment Phase:
   - Identify the specific challenges posed by the scenario.
   - Document inefficiencies, risks, or bottlenecks.
   - Example: "The software update caused delays in processing customer requests due to unfamiliar features."

2. Planning Phase:
   - Brainstorm potential solutions or adjustments to address the identified issues.
   - Prioritize solutions based on feasibility and expected impact.
   - Example: "Provide training sessions for staff on the new software interface."

3. Implementation Phase:
   - Choose one or more solutions and simulate applying them.
   - Outline steps for implementing the change.
   - Example: "Schedule two one-hour training sessions and assign a tech lead for troubleshooting."

4. Monitoring Phase:
   - Simulate tracking the results of the implemented changes using hypothetical metrics (e.g., process time, customer satisfaction scores).
   - Example: "Track how long it takes staff to process requests before and after training."

5. Feedback & Analysis Phase:
   - Gather feedback from stakeholders or team members about the changes.
   - Identify what worked well and areas for improvement.
   - Example: "Staff report improved confidence, but additional FAQs are needed for quick reference."

6. Optimization Phase:
   - Apply refinements based on the feedback.
   - Document the updated process for future reference.
   - Example: "Create a digital quick-reference guide for the software and

schedule refresher sessions."

Part 3: Debrief and Reflect

1. Review the Adaptation:
   - What adjustments were made, and how effective were they?
   - What lessons were learned from the simulation?

2. Analyze the Service Wheel Process:
   - How did each phase of the Service Wheel contribute to the solution?
   - Were there any phases that could be improved or expanded?

3. Prepare for Real-World Application:
   - How can the principles of adaptability and the Service Wheel be applied to actual challenges in your role or organization?

Evaluation Criteria:

Participants should evaluate their success based on:

- Problem Identification: Were the gaps or inefficiencies accurately identified?
- Effective Planning: Were the proposed solutions feasible and targeted?
- Measurable Results: Did the changes result in measurable improvements?
- Stakeholder Feedback: Was feedback incorporated effectively to optimize the process?

Output:

Participants should document:

- The initial process challenges.
- The steps taken in each Service Wheel phase.
- The final optimized process.
- Key takeaways for applying adaptability in future scenarios.
- By the end of this exercise, participants will have a hands-on understanding of how to adapt processes dynamically, ensuring continuous improvement and resilience in a changing environment.

PRINCIPLE 5 – CATALYST

Building on the concept of adaptability, it is imperative to recognize that every action within Service Management must be purposeful, driving meaningful progress and creating real "needle movement." In this context, a catalyst functions as the fundamental driving force behind process development and improvement, clearly identifying the reasons for change and highlighting the specific causes that necessitate action. A deep understanding of what triggers the need for change is essential, as it ensures that processes evolve effectively and remain responsive to both internal dynamics and external pressures.

Key concepts that underpin this principle include iterative change, which emphasizes the importance of adjusting in manageable, incremental steps rather than attempting sweeping transformations all at once. This approach not only mitigates risk but also allows for continuous evaluation and refinement based on real-time feedback. Equally critical is practicality, which focuses on designing solutions that are realistic, achievable, and directly aligned with the current constraints and capabilities of the organization. Additionally, visibility plays a vital role by ensuring execution of every change with complete transparency, thereby fostering trust, collaboration, and accountability among all stakeholders. Finally, continuous and deliberate reviews are essential to maintain alignment with overarching goals and to ensure that processes remain adaptable in the face of new challenges.

Deliberate Considerations (DC)

Within the Shift Service Improvement Methodology, the concept of a catalyst is supported by a set of guidelines that work in concert to guarantee that changes are not only initiated effectively but are also sustainable and impactful. These considerations—iterative change, practicality, deliberate assessments, and visibility—are interdependent and collectively ensure that every modification is well-integrated into the broader service ecosystem. Iterative change underscores the necessity of gradual progress, allowing teams to make small, strategic adjustments that cumulatively lead to significant improvements. Practicality ensures that each solution is both functional and directly contributory to the desired outcomes, balancing innovative ideas with the practical realities of the organizational environment.

Deliberate assessments are the systematic evaluations of existing processes; they help identify critical areas for improvement and prioritize changes based on their potential impact. Such assessments provide a clear snapshot of the current state and form the basis for informed decision-making. Meanwhile, visibility guarantees that all stakeholders have unfettered access to detailed information regarding the catalyst, its implications, and the planned adjustments. This open documentation process fosters a culture of transparency, enabling teams to build trust and work collaboratively towards common objectives.

By addressing the catalyst through these integrated Deliberate Considerations, Shift transforms the initial spark of change into a comprehensive, well-executed process of continuous improvement. This approach not only maximizes value but also ensures that every change is in harmonious alignment with the broader goals of the service ecosystem, driving sustained excellence and long-term success.

●●●

DC 21 - Iterative Change

Iterative change is a dynamic approach that emphasizes proactive action and adaptive responsiveness over exhaustive, long-term planning. Unlike traditional methods that depend heavily on upfront strategizing and comprehensive blueprints, iterative change focuses on making small, incremental improvements through repeated cycles. This approach is particularly valuable when the overall vision or end goal is not yet fully defined, as it allows for continuous refinement based on real-time feedback and emerging insights.

In rapidly evolving environments or situations where customer requirements shift swiftly, iterative change proves indispensable. By deconstructing complex processes into manageable, smaller cycles, teams can implement modifications gradually and adjust their strategies as needed. This inherent flexibility ensures that progress is maintained even amid uncertainty or changing demands. Instead of devoting excessive time to detailed, long-range planning, iterative change encourages teams to take immediate, actionable steps while learning continuously from each cycle. Every iteration provides an opportunity to assess what is working, identify areas for improvement, and implement necessary adjustments. This ongoing cycle of testing, evaluating, and refining helps align solutions more closely with actual requirements and user feedback.

The iterative approach is especially effective in complex or unpredictable environments where meticulous planning may be impractical. By embracing incremental change, organizations can steadily progress towards their goals, respond swiftly to unforeseen challenges, and achieve the desired cumulative outcome through a series of small, yet impactful, adjustments.

Consider, for example, the iterative production of a painting. Initially, an outline is drawn, followed by the application of basic colors, and finally, the addition of intricate details. Although each of these steps is executed once per iteration, the process is repeatable, allowing for the rapid integration of improvements, prompt error correction, and significant risk reduction. In contrast, an incremental approach involves developing each component of the painting as a discrete element, allowing for granular feedback and targeted refinement. Often, the most effective strategy combines both iterative and incremental methods, capitalizing on the strengths of each while mitigating their respective risks.

• • •

The essence of iterative change lies in its capacity to convert uncertainty into opportunity. By breaking complex challenges into manageable steps, organizations foster a culture of agile innovation and sustained progress.

Below is an illustrative diagram that outlines the process for iterative change:

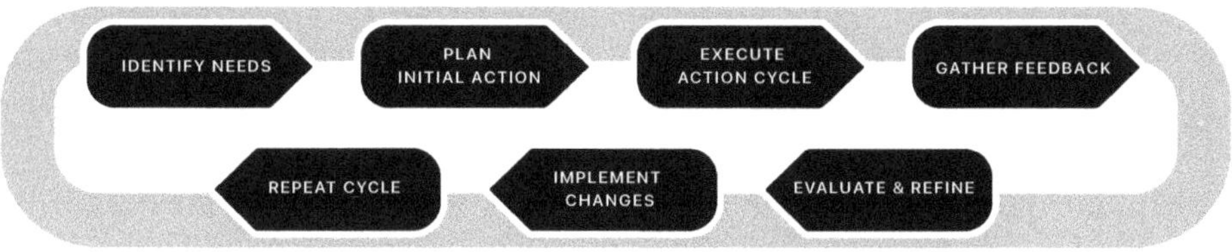

Process for Iterative Change

This diagram depicts the iterative change process: beginning with the identification of needs, followed by planning an initial action, executing the action cycle, gathering feedback, evaluating and refining the process, implementing the changes, and then repeating the cycle. This continuous loop ensures ongoing improvements and adaptation to new challenges, driving enhanced performance and innovation.

Iterative change is a flexible, feedback-driven approach that emphasizes taking small, actionable steps over rigid long-term planning. By enabling continuous learning and adjustment, it empowers teams to adapt quickly to evolving needs, especially in uncertain or fast-changing environments. This method fosters steady progress, reduces risk, and aligns outcomes more closely with real-world demands through repeated cycles of refinement and improvement.

• • •

DC 22 - Practicality

Practicality serves as a powerful catalyst for change within any process. Its effectiveness hinges not only on how useful the individual steps of a business practice are perceived, but also on the actual value generated by the resources invested. When elements of a process lack practicality, team members quickly recognize these shortcomings and may become frustrated, especially when such inefficiencies hinder their ability to deliver high-quality services. Customers, too, tend to appreciate a practical approach; when they experience positive, tangible results, they are more likely to trust and remain loyal to the service. Therefore, it is crucial to consider both the perceived usefulness and the realized potential of any process when evaluating its underlying logic.

In some instances, customers may feel that a service is not being delivered as intended, particularly when the desired objectives have not been clearly established or rigorously verified. Such discrepancies between perception and reality are almost inevitable, especially when those executing the work or receiving the service begin to question its practicality. Shift Practitioners must be vigilant in recognizing these moments, rigorously analyzing the clarity of both processes and communication. With thorough consideration, they can then develop and implement appropriate countermeasures to realign expectations and outcomes.

It is also important to note that customer requirements may not always perfectly align with the most pragmatic approach. Nevertheless, open-minded stakeholders are generally receptive to exploring alternative options for change—provided these alternatives offer tangible benefits and are not constrained by legal or regulatory mandates.

One of the most effective strategies for identifying areas that could benefit from a more practical approach is to conduct a comprehensive walkthrough of the process from start to finish, involving both the service team and customers. Such a walkthrough can be executed as a live demonstration or even as a virtual tabletop exercise, thereby conserving resources while still creating a clear narrative for efficiency. By meticulously communicating each step and documenting the actual results, teams are better equipped to pinpoint specific issues that require a more practical resolution.

•••

A thorough process walkthrough, involving both internal teams and customers, is an invaluable technique for identifying inefficiencies and uncovering opportunities for improvement. This collaborative approach not only enhances clarity but also fosters a culture of continuous, practical improvement.

By embracing practicality as a guiding principle, organizations can ensure that their processes are not only theoretically sound but also operationally efficient. This balanced evaluation—considering both perception and reality—lays the groundwork for developing processes that are resilient, effective, and capable of delivering consistent, high-quality results.

Another key to applying practicality to a function is to eliminate assumptions that have been institutionalized through repetition. Dynamic service environments will collectively endeavor to reduce thoughts of "the way that we've always done things" in team members. Customers should also expect to benefit from increased efficiency realized through the reduction of waste.

• • •

DC 23 - Deliberate Assessments

This Shift concept is fundamentally predicated on the capacity to change mindsets and maintain a relentless focus on continuous improvement. Consistent and intentional evaluation of services and their associated processes is essential for refining and enhancing performance over time. Scheduled evaluations play a critical role in ensuring that systems remain aligned with organizational goals and do not deviate into inefficiency. In particular, remnants of outdated processes—those lingering practices that persist from a decade ago—can inject confusion into an already data-rich environment, impeding progress and clarity.

Overburdening workloads, often a result of adhering to "the same way we did 10 years ago," serve to lock organizations into patterns of non-progress. It is therefore imperative to assess these antiquated practices and either automate, eliminate, or redesign steps to uncover efficiencies that were previously unattainable. There is always an opportunity for improvement, and by deliberately applying this mindset to the service environment, organizations naturally create fertile ground for further enhancements.

Complacency is the relentless adversary of progress. Efforts to counteract a complacent culture must be consistent and deliberate. Observable positive adjustments and proactive initiatives should be recognized and rewarded, as this encourages a culture of ongoing improvement. Even in the absence of obvious problems, it remains essential to evaluate processes periodically. In these instances, a deliberate, methodical assessment—whether formal, such as using a checklist, or informal, such as observational reviews—can reveal subtle inefficiencies and areas for enhancement.

Without a culture that champions superior results through iterative advancements in service delivery, established norms can become deeply entrenched, leading employees to resist change regardless of its potential benefits. Therefore, while proficiency in daily activities is important, it is equally crucial for teams to embrace regular process pulse checks. During these assessments, it is vital to consistently:

- Praise initiatives aimed at evaluating and improving processes.
- Reward active participation in process improvement efforts.
- Integrate all components seamlessly with a commitment to continuous improvement.

• • •

Regular, deliberate assessments are the linchpin of sustained improvement. By systematically evaluating both visible and subtle aspects of processes, organizations can preempt stagnation and foster an environment where innovation is not only encouraged but becomes an integral part of the daily workflow.

By embedding a culture of continuous evaluation and iterative improvement, organizations not only keep their processes aligned with evolving objectives but also empower their teams to achieve a higher level of operational excellence.

Exercise: Practicing Deliberate Assessments

Identify a process or workflow in your current work environment that has not changed much in recent years. It can be something you do daily or weekly, such as a reporting system, communication procedure, or task completion method. Write down the main steps of this process.

Assess the Process by using the following questions:

- Is there a step that feels outdated or unnecessarily time-consuming?
- Are there redundancies or bottlenecks?
- Could any part of the process be automated with tools available today?
- Does this process align with current business needs and technologies?

Note all opportunities for improvement, no matter how minor they may seem. Share your process and assessment with a partner or someone on your team. Discuss how the process has changed over time and where you see opportunities for improvement. Encourage feedback on how they might streamline or modernize the process.

Create an action plan to improve the process. Decide on one step you will take to either automate, eliminate, or simplify a part of the process. Then, commit to revisiting the process in the future to measure improvements.

Key Points to Remember:

- Praise efforts to assess and evaluate processes.
- Reward participation and initiative in process improvement.
- Aim for continuous improvement in all workflows, no matter how minor they seem.

This exercise encourages constant process evaluation, helping to avoid complacency and ensuring efficiency in an ever-evolving work environment.

●●●

DC 24 - Visibility/Transparency

Processes must be meticulously documented and made readily accessible to every individual who requires reference to them. Such documentation should serve as a comprehensive roadmap of the information repository—for example, including a detailed file plan—and must clearly define the roles and positions involved in each process. By explicitly specifying who is responsible for which tasks, as well as identifying the appropriate contacts for various issues, organizations ensure that every team member is fully aware of their collaborators and has a clear channel of communication. This level of clarity is indispensable for maintaining smooth operations and fostering effective teamwork.

In addition, thoroughly document and transparently communicate any catalysts for process improvement, such as identified shortfalls or the introduction of untapped resources. When a shortfall or a new requirement arises that necessitates a modification in the process, it is critical to promptly share this information with all relevant stakeholders. Such transparency not only facilitates the swift resolution of emerging issues but also ensures that the potential benefits of newly introduced resources are fully realized. Stakeholders, equipped with this up-to-date information, can adjust their strategies and operations accordingly, thereby contributing to overall process improvement and enhanced operational efficiency.

Furthermore, all data related to the process must be made visible and accessible to interested parties. This includes not only the raw data collected from various sources but also any modifications or filters that have been applied, along with the final outputs or products. Maintaining such transparency in data handling is fundamental to building and sustaining trust among stakeholders, as it provides a clear understanding of the data's origins, the transformations it undergoes, and its final results. This clarity is crucial for verifying the accuracy and reliability of the information and for ensuring that all parties can make informed decisions.

Additionally, the methodologies employed in the process should be clearly defined and made available when needed. By documenting and disseminating these methodologies, organizations provide a comprehensive understanding of the processes and practices in use, which in turn facilitates effective engagement with the process. This level of transparency not only improves the quality of outcomes but also significantly enhances overall operational efficiency by enabling continuous improvement across the board.

STEP	DESCRIPTION	STAKEHOLDERS	ACTION REQUIRED	VISIBILITY REQUIREMENT	TRACKING & DOCUMENTATION
DOCUMENT PROCESS	Detail every step of the process from start to finish, including resources and tools used.	Process owners, team members	Ensure all steps are clearly outlined, include roles and tools used in each phase.	Publish in a shared location accessible to all stakeholders (e.g., intranet, shared drive).	Record each process and update with new changes, ensuring all versions are stored.
DEFINE PEOPLE/ POSITIONS	Identify who is involved in the process and their specific roles.	Process owners, team members, stakeholders	List names, titles, and contact information in the documented process.	Ensure roles and contacts are included in all process documentation.	Track role updates, ensuring contact information is current.
IDENTIFY CATALYSTS FOR CHANGE	Highlight any shortfalls or new resources that trigger changes in the process.	Process owners, improvement teams, management	Monitor for shortfalls or resources and update process documentation accordingly.	Ensure change catalysts are visible and accessible to all impacted parties.	Track identified shortfalls and resources, ensuring stakeholders are notified.
COMMUNICATE SHORTFALLS	Make all stakeholders aware of any gaps in the process that need addressing.	Process owners, stakeholders	Communicate shortfalls through appropriate channels (email, meetings, platforms).	Make shortfall data visible to all stakeholders to enable collective problem-solving.	Maintain logs of shortfalls and actions taken to address them.
ANNOUNCE NEW RESOURCES	Ensure all stakeholders are aware of any new resources introduced to the process.	Process owners, improvement teams, all stakeholders	Communicate the availability and purpose of new resources.	Publish new resources and make them available to all relevant stakeholders.	Track resource announcements and their impact on processes.
DATA VISIBILITY	Make all collected data, applied edits/filters, and outputs visible to relevant stakeholders.	Data analysts, stakeholders, customers	Maintain transparency in data collection and outputs, enduring full visibility.	Provide access to all data in its raw and process forms.	Track data usage, edits, and final products, with clear documentation of methods used.
METHODOLOGY TRANSPARENCY	Define and make visible the methodologies used for collecting and analyzing data.	Data analysts, stakeholders, auditors, customers	Document all methodologies used in the process.	Make methodologies visible to those requesting or impacted by the data outcomes.	Maintain detailed records of methodologies and update for auditing purposes.

The outlined steps for effective process management begin with documenting the entire process, detailing every step from start to finish, along with the resources and tools used. This task involves process owners and team members, who must strive to ensure clarity in each phase and publish the documentation in a shared location accessible to stakeholders. It is essential to define the people and positions involved, listing names, titles, and contact information to facilitate communication. Include this information in process documentation, with updates tracked to ensure accuracy.

Identifying catalysts for change is the third step, where process owners, improvement teams, and management must highlight any shortfalls or added resources prompting changes. Monitor collected information and make it visible to affected parties. Communicating shortfalls is the fourth step, which involves notifying stakeholders of gaps that require attention through appropriate channels. Maintaining logs of these shortfalls and actions taken is essential for accountability. During the fifth step, inform stakeholders of any added resources introduced to the process, with announcements tracked to assess their impact.

The sixth step emphasizes data visibility, where data analysts and stakeholders must ensure collected data, edits, and outputs are transparent and accessible. Tracking data usage and methods is vital for clarity. Finally, the last step focuses on methodology transparency, requiring data analysts to document the methods used for data collection and analysis. Make this information available to relevant stakeholders, with detailed records maintained for auditing purposes. Together, these steps create a comprehensive framework for process management that promotes accountability, transparency, and continuous improvement.

Summary of Visibility/Transparency Workflow

- Process Documentation: Ensure all processes and steps are fully documented and stored in a transparent location.
- Define Roles: Clearly define the people involved and their roles in the process, making this information accessible.
- Catalysts for Change: Identify any shortfalls or added resources and ensure they are communicated to stakeholders.
- Data Transparency: Ensure data, from input to output, remains transparent and accessible.
- Methodology Visibility: Document and make methodologies used in processes available to anyone affected by or interested in the data.

This workflow ensures that processes, data, and improvements are visible and accessible to everyone involved, fostering accountability and collaboration.

Effective process management requires thorough documentation, clear role definition, and open accessibility to ensure smooth operations and cohesive teamwork. Transparency around changes—such as shortfalls or new resources—and visibility into data and methodologies empower stakeholders to adapt strategies, verify accuracy, and drive continuous improvement. This structured, transparent approach builds trust, enhances accountability, and supports long-term operational efficiency across the organization.

PRINCIPLE 6 – TARGETS

Every process must operate with a clearly defined target, establishing a desired outcome that serves as the guiding beacon for all subsequent activities. Maintaining unwavering focus on this target ensures that every step taken is purposeful and directly contributes to achieving the intended results. To this end, establishing measurable indicators is essential for rigorously assessing progress toward the desired end state and ensuring that outcomes remain aligned with expectations. By systematically implementing these measurements, organizations can effectively monitor performance, make well-informed adjustments, and attain optimal results in their processes.

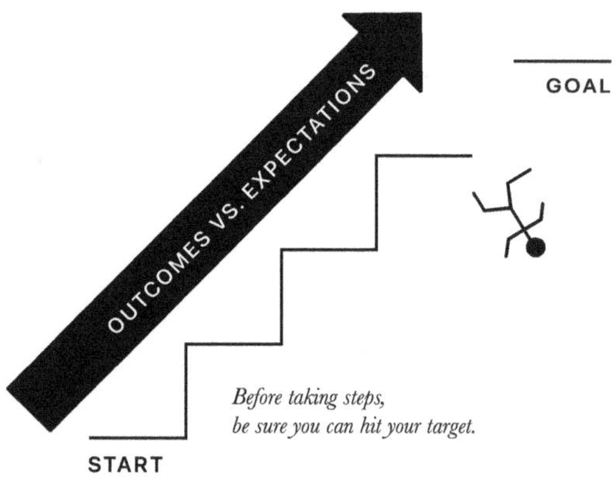

Deliberate Considerations (DC)

The Shift methodology emphasize that processes should be designed and continually improved so they can be executed repeatedly, yielding near-identical outcomes while also incorporating mechanisms for ongoing refinement. The primary focus is on establishing a robust framework that sustains consistency yet remains sufficiently flexible to adapt to changing conditions and evolving requirements. Recording outcomes and meticulously evaluating each step in the process is crucial for accurately assessing the effectiveness of service delivery and any implemented modifications. This methodical approach not only enables organizations to identify best practices but also highlights areas where further improvement is possible.

Without the establishment of clear, quantifiable measurements to gauge success, it becomes nearly impossible to validate achievements or understand the true impact of any changes. These metrics function as a strategic compass, guiding teams in their decision-making processes by providing valuable insights into performance trends over time. Moreover, they foster a culture of accountability, ensuring that every team member comprehends their role in the overall success of the process. By systematically analyzing results, organizations can make informed adjustments and continuously enhance their processes, thereby ensuring that they not only meet but eventually exceed established expectations. A dual focus on both consistency and adaptability in process design leads to improved service delivery and a higher level of overall organizational effectiveness.

Within the Shift framework, the goal is unequivocally to make everything measurable. Process improvement initiatives should yield efficiencies that are readily visible to both team members and customers alike. A typical demonstration of this visibility is performed through detailed reports that contain numerical data and specific outcomes. Furthermore, whenever possible, processes should be designed or modified in a way that preserves their repeatability and leaves room for future enhancements.

DC 25 - Enrichment

Value-add refers to the creation of supplementary elements that exceed the original worth of a product or service, often by introducing novel features or enhancements that were previously unavailable. While such an approach can undoubtedly enhance customer satisfaction by providing additional benefits, it does not always capture the deeper, more transformative essence of enrichment. Enrichment involves actively improving or refining the inherent quality of a product, service, or process to elevate the overall customer experience, rather than simply accumulating extra features or resources. This distinction is critical because merely adding more to a service does not inherently translate into a better offering—particularly when the original product already effectively meets customer needs. In some cases, superfluous additions may, in fact, complicate the user experience rather than improving it.

Enrichment is a deliberate and strategic effort to enhance a specific output, ensuring that every improvement directly benefits the customer or the organization. It focuses on substantive advancements—such as streamlining processes for greater efficiency, improving a product's usability, or tailoring a service to more precisely address customer needs—thereby delivering a deeper sense of value that resonates profoundly with users. Unlike the concept of value-add, which can sometimes prioritize quantity over quality, enrichment ensures that each modification contributes meaningfully to the overall improvement of the targeted area.

This concept is universally applicable across a broad spectrum of domains, including services, products, processes, and management practices. In the realm of Service Management, for example, enrichment might involve integrating advanced technology to streamline workflows or providing customized training to employees to increase engagement and productivity. In product development, it could mean improving durability or usability without introducing unnecessary complexity. By concentrating on an elevated level of development, organizations can forge lasting relationships with customers and stakeholders, demonstrating a steadfast commitment to not only meeting but consistently exceeding expectations through thoughtful and meaningful enhancements.

Regard enrichment as a cornerstone of continuous improvement. It necessitates a considered approach firmly grounded in a comprehensive understanding of customer expectations, organizational objectives, and the broader service ecosystem. By purposefully designing enhancements aligned with these critical elements, companies can ensure that their efforts yield tangible benefits and create a sustained impact that far surpasses the initial value of the product or service. This focus fosters a culture of excellence in which every improvement is intentional, impactful, and meticulously aligned with the goal of delivering an exceptional customer experience.

●●●

Cultivating enrichment is about quality over quantity – prioritize meaningful enhancements that truly add value rather than simply increasing the number of features. This focused approach ensures that improvements resonate with customers and drive long-term success.

MENTAL ESCAPE: ENRICHING PROCESSES

How do you enrich the processes around you?

Take a moment to picture a Service Management environment where enrichment is a key driver of success. Imagine you, as a Service Manager, constantly seeking new opportunities to improve processes, always with an eye toward innovation and growth. Now, visualize yourself engaging with key stakeholders before any initiative is implemented. Feel the energy of collaboration as everyone comes together, ready to share ideas, concerns, and solutions.

As you move through this Mental Escape, see yourself facilitating discussions where every voice matters. Picture an open space where stakeholders feel free to express their ideas and share feedback without hesitation. Together, you engage in meaningful conversations that lead to creative growth, fostering an environment of trust and innovation. In these moments, compromise is not a burden but a strength, as diverse perspectives blend into actionable strategies for improvement.

Now, redirect your focus from the discussions to the execution phase. Visualize a clear, strategic pathway that transforms those collaborative ideas into measurable outcomes. You have set concrete goals, and with the team's buy-in, you feel confident in the transformation ahead. As the enrichment process takes shape, imagine the sense of accomplishment, knowing that the improvements are not just theoretical but real, impactful changes that elevate the entire Service Management experience.

Breathe in this sense of achievement, knowing that true process development thrives on collaboration, creative freedom, and shared ownership. Let yourself feel the satisfaction of leading a team toward measurable transformation, one enrichment initiative at a time.

•••

DC 26 - Customer Service

Voice of the Customer (VoC) refers to the comprehensive feedback provided by customers regarding their experiences with, and expectations for, products or services. This concept encompasses not only customer needs and expectations but also their overall understanding of the service and suggestions for product improvement. It serves as a vital barometer for gauging the quality of customer service and identifying opportunities for enhancement.

Service Managers, at their core, aim to delight customers, and companies consistently strive to garner positive feedback. Organizations often go to great lengths to encourage customers to express favorable opinions about their services, sometimes even incentivizing participation in surveys to collect this valuable feedback. However, the true essence of VoC extends beyond simply accumulating praise; it is equally important to establish genuine, open channels of communication that allow for the collection of critical, negative feedback alongside positive comments. In the Shift methodology, this Deliberate Consideration is dedicated to creating a Customer Service environment that consistently exudes a positive atmosphere through well-structured intent and decisive action. Negative feedback, far from being disregarded, is viewed as an essential source of insights that can drive continuous improvement.

Rather than attempting to narrowly define customer service, Shift embraces a broader interpretation—providing exceptional service to every customer. In this light, VoC is not just about collecting data; it is about understanding the underlying needs and expectations of customers and using that knowledge to drive product and service enhancements. Shifting the focus to tangible measurements, the timely and accurate assessment of delivered services is paramount. In situations where direct feedback from customers is unavailable, teams must proactively evaluate their services and integrate continuous improvement initiatives. For example, if a team or individual excels in a particular task, a continual improvement register captures that success and generates ideas for further enhancements. This process of constant evaluation, systematic assessment of outputs, and rethinking of strategies is a deliberate effort and an essential component of continuous improvement.

In summary, Voice of the Customer is a multifaceted concept that plays a crucial role in Service Management. By focusing on both qualitative and quantitative aspects of customer feedback, organizations can ensure that their services continually evolve to meet and exceed customer expectations, thereby fostering an environment of ongoing improvement and excellence. To capture these valuable insights and drive continuous progress, it is essential to document and systematically track improvements through a Continuous Improvement Register (CIR), which the following section covers in detail.

The above image represents the columns for a sample CIR. The CIR is a structured

IMPROVEMENT ID	DATA LOGGED	IMPROVEMENT AREA	DESCRIPTION OF IMPROVEMENT	CATALYST/TRIGGER FOR IMPROVEMENT	ACTION OWNER	STATUS	TARGET COMPLETION DATE	OUTCOMES/ COMMENTS

document or database used to track and manage ongoing efforts to enhance processes, services, or workflows within an organization. It serves as a tool for documenting improvement ideas, monitoring their progress, and assessing their impact. Continuous improvement (CI) focuses on incremental, ongoing improvements rather than large-scale overhauls.

Key Components of a Continuous Improvement Register:

- Improvement ID: A unique identifier for each improvement initiative to ensure easy reference and tracking.
- Date Logged: The date when the improvement was first identified and added to the register.
- Improvement Area: The specific process, department, or function that the improvement effort is targeting, such as data reporting, customer service, or process documentation.
- Description of Improvement: A brief explanation of the proposed improvement, outlining what will be changed and how it will benefit the organization.
- Catalyst/Trigger for Improvement: The reason behind the proposed change, such as an observed issue, customer feedback, inefficiency, or technological advancement.
- Action Owner: The person or team responsible for implementing and overseeing the improvement. Accountability ensures on-time completion of tasks.
- Status: The current stage of the improvement initiative (e.g., "Not Started," "In Progress," "Completed"). The status helps track progress over time.
- Target Completion Date: The estimated date by which the improvement will be completed, helping teams set deadlines and prioritize tasks.
- Outcome/Comments: Space for capturing the results of the improvement effort and any observations or follow-up actions. This section provides

insights into the effectiveness of the change and whether additional action is needed.

Benefits of Using a Continuous Improvement Register:

- Visibility & Accountability: All improvement initiatives are visible in one place, ensuring that each one is tracked and assigned to the owner responsible for progress.
- Prioritization: Helps teams prioritize which improvements to focus on based on their potential impact or urgency.
- Measurement of Progress: Teams can easily see the status of each initiative, helping to maintain momentum and keep projects on track.
- Documentation for Future Reference: By keeping a log of improvements, organizations can assess past initiatives and avoid reinventing the wheel when similar challenges arise.
- Cultural Reinforcement: A CI register supports a culture of continuous improvement, encouraging employees to contribute to enhancing processes and workflows.

This register forms the backbone of a structured improvement process, helping organizations maintain focus on refining their operations and services over time.

●●●

Cultivating enrichment is about quality over quantity – prioritize meaningful enhancements that truly add value rather than simply increasing the number of features. This focused approach ensures that improvements resonate with customers and drive long-term success.

●●●

DC 27 - Outcomes

Every process yields an outcome, typically evaluated against specific goals and targets representing the intended results. These outcomes often serve as inputs for subsequent processes, such as collecting feedback or continuing within a broader workflow. However, it is essential for Shift practitioners to go beyond merely assessing whether predefined goals have been achieved. Practitioners must also carefully evaluate unintended outcomes, which can have a significant impact on the entire service ecosystem. These unintended results, whether positive or negative, often reveal critical insights that may not be immediately apparent.

Improving one area might appear successful at first glance but could inadvertently create inefficiencies, increased workloads, or disruptions in other processes. A narrow focus on achieving the intended targets may overlook these broader consequences, which could undermine overall performance. For this reason, Shift emphasizes a holistic view of processes, recognizing that no outcome occurs in isolation. By thoroughly evaluating both intended and unintended outcomes, practitioners can ensure that changes are balanced, sustainable, and beneficial across the organization.

Intended outcomes are assessed against predefined targets, requiring clear performance measurements and the flexibility to adjust when deviations occur. An iterative approach provides teams with the opportunity to refine their processes continuously, addressing issues and improving results. Incremental assessments are equally important, prioritizing elements based on their likelihood of recurrence. For example, attention should first be given to factors highly likely to repeat in the current process flow, followed by those that may recur in future or similar efforts, and finally, to elements with minimal chances of repetition.

Unintended outcomes, on the other hand, demand a unique perspective, as they often highlight unanticipated effects within the broader ecosystem. These outcomes might include unforeseen benefits or challenges, such as inefficiencies impacting peripheral stakeholders, processes, or technologies. Avoiding excessive focus on intended targets is critical, as it can obscure visibility into these unintended impacts. Iterative assessments and incremental ecosystem checks are necessary to identify and address these effects, ensuring that all elements remain aligned with organizational objectives. This balanced approach not only refines processes and improves outcomes but also promotes continuous improvement across the service ecosystem.

To ensure effective evaluation, targets should be well-defined, specific, attainable, and measurable. Comprehensive documentation should clearly outline targets and desired outcomes. Tangible results, such as meeting specific specifications or delivering measurable improvements, are often easier to assess. For instance, a service designed to meet certain specifications (e.g., x speed, y cost, z practicality) might achieve these goals and exceed them. Intangible results, while harder to quantify, can also provide valuable insights. These may include unplanned benefits, such as added features or enhancements that increase customer satisfaction. Creative measurement techniques can help capture these intangible outcomes, ensuring full consideration of both measurable and qualitative results during the evaluation process. This approach fosters a culture of accountability, adaptability, and sustained excellence.

Processes should be evaluated not only by whether they meet intended targets, but also by identifying unintended outcomes that may affect the broader service ecosystem. Improvements in one area can create inefficiencies elsewhere if not viewed holistically. Shift emphasizes an iterative, balanced approach—measuring both tangible and intangible results, prioritizing recurring patterns, and using creative evaluation techniques. This ensures process enhancements are sustainable, aligned with organizational goals, and continuously refined to support long-term success.

Exercise: Target Setting Workshop – Defining Measurable Goals

This workshop helps participants develop measurable indicators for a specific project or initiative, ensuring clarity and alignment with organizational objectives. Participants will collaborate, share insights, and refine their goals through peer feedback.

Objective:
Enable participants to set clear, measurable targets for a project, aligning them with organizational goals and ensuring they are actionable and trackable.

Part 1: Preparation

1. Select a Project or Initiative: Participants should choose a real or hypothetical project relevant to their role. Example projects:
   - Improving customer satisfaction scores.
   - Reducing operational costs in a department.
   - Implementing a new training program.
2. Understand SMART Criteria: Review the principles of SMART goals:
   - Specific: Clearly define the target.
   - Measurable: Quantify the outcome.
   - Achievable: Ensure feasibility.
   - Relevant: Align with organizational priorities.
   - Time-bound: Set a deadline for achievement.

Part 2: Workshop Activity

1. Define the Goal: Each participant writes a concise goal statement for their project.

   Example: "Increase customer satisfaction scores by 10% within six months by improving response times and service quality."
2. Break Down into Measurable Indicators: Participants identify KPIs for their goal, focusing on measurable outcomes.

   Example Indicators for Customer Satisfaction:
   - Average response time reduced from 24 hours to 12 hours.

- Customer satisfaction survey scores improved from 80% to 90%.
- The number of resolved customers' tickets increased by 15%.

3. Create a Timeline: Specify deadlines for achieving the overall goal and each KPI.

Example Timeline:

- Month 1: Analyze current response times and identify bottlenecks.
- Month 3: Implement process improvements and train staff.
- Month 6: Achieve the 10% satisfaction score increase.

Part 3: Peer Review and Feedback

1. Present Goals and Indicators: Participants share their goals and measurable indicators with a partner or small group.
2. Feedback Session: Peers provide constructive feedback, focusing on:
   - Are the goals and indicators aligned with the SMART criteria?
   - Are the KPIs specific, relevant, and actionable?
   - Are the timelines realistic and achievable?
3. Refinement: Participants revise their goals and indicators based on feedback.

Part 4: Final Review and Reflection

1. Group Discussion:
   - Share refined goals with the larger group.
   - Discuss shared challenges and solutions encountered during the exercise.
2. Reflection Questions:
   - What made setting measurable targets challenging?
   - How can this process be applied to other projects or initiatives?
   - What steps will you take to ensure effective monitoring of these targets?

Output:

Each participant will leave the workshop with:

- A clearly defined goal for their project.
- Measurable indicators and a timeline for achieving them.
- Actionable insights from peer feedback to enhance their approach.

This exercise ensures participants understand the importance of setting measurable targets and equips them with a structured method for defining and tracking project goals effectively.

●●●

THE TECHNOLOGY TENET

PRO in the Loop

Purpose of technology in the ecosystem, Requirements for intelligent design, and Outcomes that measure transformative exponential growth.

Technology in the Shift framework embodies the PRO in the Loop concept, an approach that acknowledges the critical role of human-in-the-loop machine learning. This concept underscores the idea that optimal outcomes are achieved only when human expertise and technological capabilities are seamlessly integrated. Rather than viewing technology as a standalone solution, this model emphasizes the deliberate evaluation of which functions technology should and should not perform. By striking this balance, organizations can ensure that technology serves as an enabler rather than a replacement, enriching the Service Management landscape through a continuous feedback loop that refines both technological and human contributions.

At its core, technology represents the practical application of conceptual, scientific, or technical knowledge to achieve specific objectives. Within the Shift environment, effectively leveraging technology requires more than just implementing tools—it necessitates harnessing the collective intelligence, experience, and creativity of the individuals who interact with these systems. True technological integration harmonizes processes that either support or directly utilize technology while simultaneously amplifying human potential to drive innovation. By thoughtfully aligning technological advancements with human oversight, organizations can enhance service delivery, streamline operations, and foster a culture of continuous improvement.

The PRO in the Loop concept reinforces the importance of maintaining a balanced perspective on technological innovation. While automation and digital transformation can significantly enhance efficiency, human insight remains indispensable in areas requiring judgment, adaptability, and strategic thinking. This synergy ensures that organizations remain agile and responsive, capable of evolving alongside emerging technologies without losing sight of the critical role that human expertise plays in decision-making and service optimization. By prioritizing this integrated approach, organizations cultivate a Service Management strategy that is both forward-thinking and resilient, ensuring sustained success in an increasingly complex technological landscape.

PRINCIPLE 7 – PURPOSE

Technology functions as a multi-purpose tool, offering a vast array of solutions that can be applied across numerous contexts. Successfully implementing a technology-centric approach requires identifying the right tool for the specific function at hand. Just as a multi-tool provides various implements for different tasks, selecting the correct technology for a given need is crucial to ensuring efficiency and effectiveness. While a knife on a multi-tool can technically turn a screw, a screwdriver is far better suited for the task. Similarly, the intended purpose of a technological solution may not always be immediately apparent, necessitating a thorough evaluation before implementation to ensure that the chosen technology aligns with both operational needs and strategic objectives.

Deliberate Considerations (DC)

The Purpose Principle is central to determining what is required within an environment to achieve a specific desired outcome. Effective Service Managers must position themselves as trusted advisors, providing well-informed recommendations to their customers and partners. These proposed solutions may range from high-tech to low-tech, but practicality should always be a primary consideration. Program management demands a diverse skill set, requiring professionals to perform in multiple roles, all while maintaining a focus on customer needs. Listening closely to customers provides valuable insights into their requirements, guiding the selection of the most suitable technological solutions. While different pathways may lead to the desired outcome, navigating these routes—including overcoming potential roadblocks—must be tailored to fit the specific demands of the service environment.

A well-defined purpose serves as the guiding principle for evaluating and selecting technologies, ensuring that the chosen solution meets the fundamental requirement while also delivering added value. Achieving this purpose does not necessarily require costly, cutting-edge innovations; often, low-cost solutions, existing capabilities, or incremental advancements are sufficient to meet objectives. In many cases, leveraging available technologies can provide an immediate solution, with additional enhancements and refinements introduced over time as new tools and capabilities emerge. By prioritizing practicality, adaptability, and strategic alignment, organizations can implement technology in a way that is both efficient and scalable, ensuring long-term success in a rapidly evolving digital landscape.

Technology, like a multi-tool, offers versatile solutions—but effectiveness depends on choosing the right tool for the job. Purpose-driven decision-making is essential for evaluating which technologies align with operational needs, strategic goals, and the customer environment. By grounding Service Management in clearly defined purpose statements, organizations ensure practical, scalable, and human-centered implementation of technology.

To kick off this section, the *"Putting Purpose to Work for Technology"* exercise challenges teams to articulate clear purpose statements for their tasks—sharpening focus on what truly matters and revealing how even well-intentioned solutions must be examined in context to avoid unintended consequences.

Exercise: Putting Purpose to Work for Technology

Practice writing purpose statements for any task you or your team performs. Here are some examples of purpose statements for some sample companies:

Company A –
Industry: Cybersecurity & Risk Management
Purpose Statement: To safeguard digital ecosystems through adaptive, AI-driven security frameworks that empower organizations to operate confidently in a connected world.

Company B –
Industry: Sustainable Agriculture & Food Tech
Purpose Statement: To close the loop between farms and cities by delivering hyper-local, zero-waste food systems that nourish communities and regenerate the planet.

Company C –
Industry: Digital Content & Immersive Storytelling
Purpose Statement: To spark imagination and engagement by blending narrative design with emerging technologies, creating unforgettable experiences across platforms.

Are those purpose statements good or bad, how would you rank them and why?

In application, defining purpose in technology is the guiding light that can help translate customer needs and preferences into a path forward while inducing strategic change. As an example, looking at the Company A purpose (also the mission), the company devotes considerable effort to evaluate accessibility in their technology. That mission ties directly to human progress, communications, and entertainment. Accessibility itself does not require a change to the existing experience, and it does not cost a lot to enable voice dictation. It also does not require new equipment or specific equipment to access the full feature set. Accessibility tracks well to purpose without taking away from other capabilities and initiatives within the company.

Purpose is crucial to evaluating the existing technologies and future capabilities within mission sets. Collaboration with key stakeholders, along with research and development are tools necessary for addressing technology. Through collaborative environments, customers can evaluate technologies from differing perspectives to ensure development and implementation of a well-rounded solution. Purpose keeps collaborative teams concentrating on the identity of the customer and the ecosystem; without a mindful approach to tracing everything back to purpose, technological innovation can either help advance an ecosystem, change it entirely, or cause its demise.

Concentrating on purpose should drive attention to the bigger picture. Focusing too much on a single idea, priority, or solution can cause people to lose sight of the bigger picture. Correcting one problem or improving a process could have a ripple effect on other areas and an implemented solution, no matter how ideal it seems, might adversely impact another business area or resource.

●●●

DC 28 - Technology-Driven Advantage

The Technology-Driven Advantage represents the distinct benefits created when technology improves the functionality and utility of services, processes, products, or management practices. This advantage goes beyond meeting basic needs, driving measurable improvements that directly contribute to satisfying end customers. It manifests in areas such as improved processes, time savings, cost efficiency, innovation, and human resource optimization. For technology to create a meaningful advantage, organizations must continually identify opportunities to refine its application, engage key stakeholders, and foster collaborative efforts that encourage creativity and adaptability.

Technology, when thoughtfully integrated into an environment, should offer clear and measurable benefits to existing tools and processes. It should facilitate smoother communication, enhance productivity, and enable creative problem-solving. For example, technology can streamline communication by replacing outdated methods with real-time messaging platforms or virtual meeting solutions, allowing teams to work more effectively across diverse locations. Similarly, it can boost productivity by enabling teams to collaborate in dynamic environments, combining inputs from multiple contributors to achieve single tasks efficiently. In terms of creativity, technology opens pathways for individuals to reimagine how tasks are performed, often achieving results with greater precision and less effort.

From strategy to execution, Service Managers can confidently achieve measurable transformation by leveraging the Technology-Driven Advantage. This begins with analyzing the effects and capabilities of a chosen technology. The impact of a technological change must align closely with its intended purpose, ensuring that its outputs consistently support the organization's goals. A cyclical approach to evaluating these effects—through measuring outputs and iterating based on data—helps fine-tune the technology's role in achieving desired results. When goals and technological effects align, the organization gains a powerful new capability that elevates operational efficiency and supports mission success.

To apply the Technology-Driven Advantage effectively, consider a recent technological implementation within your organization. Reflect on how technology enhanced existing processes, improved communication, or increased productivity. Assess the strategies employed to ensure these improvements were measurable and sustainable, including involving key stakeholders and incorporating feedback. Finally, think about how the technology's impact was tracked over time to ensure it continued to align with evolving organizational goals. By maintaining a focus on adaptability and purposeful measurement, the Technology-Driven Advantage can drive continuous improvement and meaningful transformation.

This arises when technology is purposefully integrated to enhance services, processes, and outcomes in measurable ways—such as improving efficiency, reducing costs, and boosting innovation. To realize this advantage, organizations must align technological solutions with strategic goals, engage stakeholders, and continuously evaluate impact through data-driven iteration. When done effectively, technology becomes a catalyst for sustained transformation, empowering teams to deliver smarter, faster, and more adaptive results.

MENTAL ESCAPE: INFUSING TECHNOLOGY INTO YOUR WORK ENVIRONMENT

Take a moment to reflect on your current work environment. Think about the technologies you use daily—are there any that have inadvertently created obstacles instead of streamlining your tasks? For instance, consider the introduction of an office coffee machine. While it might have promised convenience, does it now complicate your routine, making it challenging to grab a quick cup of coffee? Perhaps it requires multiple steps or has a confusing interface, causing you to waste precious time instead of enjoying your caffeine fix.

Now, shift your focus to your home environment. Do you have a smart device network that was supposed to enhance your daily life but instead complicates it? You may find that operating your sound system has become a cumbersome task, leaving you sitting in silence instead of enjoying your favorite music.

But let us dig deeper. Is there a more serious technology issue impacting your day-to-day tasks? Perhaps it is a cumbersome project management tool that complicates rather than simplifies collaboration, leading to frustration and confusion among your team. Such situations can lead to diminished productivity and increased stress. Take a deep breath and visualize a more efficient work environment. Imagine a space where technology seamlessly integrates into your daily routine, enhancing your productivity rather than hindering it. Picture an office where the coffee machine is intuitive, allowing you to enjoy your beverage quickly, and where smart devices work harmoniously.

As you envision this ideal scenario, consider the steps you could take to address the technological challenges you currently face. It could be time to speak with your team about the coffee machine's usability or explore alternative project management tools that better fit your workflow. Embrace the idea of advocating for solutions that will enhance your work and home life. By taking proactive measures to address technology-related frustrations, you can foster a more productive, enjoyable, and efficient environment, improving your overall well-being and job satisfaction.

●●●

DC 29 - Collaboration

Technology has revolutionized collaboration, providing powerful tools to connect individuals, share knowledge, and facilitate collective problem-solving. Platforms like Teams, Slack, and Google Workspace enable real-time communication, resource sharing, and streamlined program management, making collaboration more efficient and accessible. These technologies allow diverse stakeholders to contribute meaningfully, regardless of location, ensuring that every voice is heard and every perspective considered. By providing a robust infrastructure for pooling skills and resources, these tools create transparency and inclusiveness in collaborative efforts.

Despite these advantages, technology introduces unique challenges. The ease of access to information and tools can create an illusion of self-sufficiency, leading individuals to believe they no longer need others for assistance. This misconception can erode the human aspects of collaboration, such as trust, empathy, and the creative constructive collaboration that arises in group settings. Over-reliance on technology may also result in miscommunication due to the absence of non-verbal cues or an overemphasis on the tools themselves rather than on fostering relationships and meaningful interactions.

To effectively leverage technology for collaboration, it is essential to balance its capabilities with the foundational principles of teamwork. Utilize technology to enhance engagement and inclusivity, not replace human connection. Organizations can achieve this balance by intentionally designing workflows that encourage stakeholder input, fostering open dialogue, and providing training to ensure all team members can utilize collaborative tools effectively. In purpose-driven work, the goal is not just efficiency but achieving impactful and well-rounded solutions. By linking the advantages of technology with efforts to preserve the human elements of teamwork, organizations can create deeply connected and highly productive collaborative environments. This synergy transforms individual contributions into collective success while mitigating the risks of over-reliance on technology.

A significant challenge in collaborative settings is overcoming cognitive bias, the tendency to make decisions based on individual experiences, preferences, or perceptions that may not align with objective reality. Cognitive bias can lead to overconfidence in a particular approach, resulting in decisions that fail to address the core issue or meet broader organizational goals. To counteract this, Service Managers must view situations from multiple perspectives and ensure the inclusion of all relevant stakeholders in the conversation. Bringing in an outsider or someone with an unconventional viewpoint can challenge assumptions and encourage the group to reconsider its approach, preventing groupthink and ensuring well-considered solutions.

Maintaining a detailed stakeholder list is another essential strategy for effective collaboration. Keeping track of key players ensures inclusivity and prevents decisions from being dominated by the most vocal participants. A data-driven approach helps Service Managers verify representation of relevant perspectives, avoiding the pitfall of limiting collaboration to a narrow group. This ensures that decisions are based on a comprehensive understanding of the situation and align with organizational objectives.

Group dynamics also play a critical role in the success of collaboration. Strong dynamics and high energy can sometimes lead to overconfidence and missed opportunities, while weak dynamics may cause stagnation. In cases where group energy is low, skilled facilitation can help re-energize the team, improve communication, and refocus efforts. Conversely, introducing an outlier in high-energy groups can ground discussions, bringing overlooked alternatives to the table and broadening the range of viable solutions.

Effective collaboration requires more than bringing people together; it demands an environment that values diverse perspectives, and the minimization of cognitive biases. By ensuring inclusivity, maintaining a focus on data-driven processes, and managing group dynamics thoughtfully, organizations can foster a culture of purposeful collaboration. This intentional approach drives innovation and leads to more successful outcomes, ensuring that collaboration evolves to meet the needs of the group and the mission at hand.

MENTAL ESCAPE: EMBRACE THE CRITIC OR STAY IN YOUR COMFORT ZONE

Picture yourself in the middle of a lively group discussion. Feel the energy of voices chiming in, ideas flowing back and forth. Now, ask yourself, do you prefer the comfort of consensus, where everyone nods in agreement, or do you secretly thrive when someone challenges your ideas? Are you the type who embraces the critic, eagerly inviting feedback, or do you gravitate toward those who affirm your perspective, seeking reassurance in familiar opinions?

Now, think about the decisions you have made lately. Have you ventured outside your usual circles to gather fresh perspectives, or have you stuck with the opinions that feel safe? Sometimes, the most valuable insights come from the unexpected—the voices that question, probe, and challenge. Imagine how your thinking could expand by stepping into discomfort, into the space where new ideas take shape from constructive friction.

Reflect on the people in your life: who always cheers you on, supporting your ideas without hesitation? And who challenges you, making you work harder to refine and prove your point? What could you gain by spending more time with the critics, the ones who push you beyond your comfort zone?

Now, take a moment to imagine your ideas evolving in this space of tension. Envision how welcoming more criticism could open new paths, spark innovations, and lead you to exciting breakthroughs. Sometimes, it's not the easy road that leads to the most rewarding destinations—it's the one paved with challenges, where growth happens, and ideas truly come to life. Allow yourself to embrace the discomfort and consider how your creativity and resilience might flourish with a bit more friction in the mix.

• • •

DC 30 - Research and Development

Research and development (R&D) is comprised of creative and systematic work undertaken to increase intellectual capital and to devise new applications of available knowledge. When defining R&D as applied to the Purpose principle, convey the value that it can have on a project, as it generates pathways of creativity and exploration with a methodical approach to identifying new ways to achieve goals. When used appropriately, engineering teams can offer purpose-driven solutions or outside-the-box ideas. Shift emphasizes the Service Management methodology that allows R&D processes to occur while facilitating and invigorating creativity. Therefore, Shift Service Practitioners should endeavor to examine the aspects and tendencies that will have the most negative impact on what would otherwise be a highly productive effort. Micromanagement, lack of autonomy, ambiguous objectives, and promoting a siloed work environment are problematic tendencies within an R&D effort. Prevention of those inhibiting factors will certainly cultivate a service environment postured for success when introducing new technology into the ecosystem.

Hierarchical management paves a road to the centralization of power. Typically, this leadership structure introduces a micro-management scenario which innately funnels research and development through one focal point. Although micromanagement can be a successful approach in certain scenarios or applications of management, it will likely work against the creative nature often found in R&D teams. It can lead to disempowered teams and often leads to individuals within the team feeling replaceable. Furthermore, these teams are frequently comprised of engineers who often pride themselves on autonomy, an opposing force against managers that prefer monitoring and controlling the daily activities of their employees.

Another problematic area to avoid is applying skills and resources toward ambiguous objectives. Advancing that concept a step further, R&D teams also tend to start working toward a defined objective and then slip off target. This loss of focus can happen when research occurs and leads a team down a path to create something that appears to meet objectives. By the time efforts are evaluated, the project takes on an entirely different look than when the team first began working on a task. Before too long, stakeholders begin to disassociate themselves from the project because they may have realized the undesired evolution but waited too long to speak up. Which can lead to the last negative tendency to avoid, the presence of siloed work environments.

Siloed work environments occur elsewhere in the service ecosystem and cause teams to stop communicating efficiently and effectively. These "silos" can also cause information hoarding and partitioned workloads, leading to severe inefficiencies within the service ecosystem. Those inefficiencies are represented as duplication of tasks and wasted resources. The breakdowns in communication will erode coordination between the teams and slow decision-making processes, leading to a misalignment in goals. These things can tear apart any work environment, but within R&D functions, they counteract nearly every task within the function.

After addressing what to avoid, the next step is leveraging capabilities to drive technology in an organization. Productive, well-rounded teams are solution oriented. While R&D has its own guiding principles, a Service Manager can use Shift principles to target specific technology applications. This approach does not just lean forward, but does so with an eye on the mission, resources, environmental factors, and stakeholder appetite for change. Service Managers should first utilize existing resources like software and systems to enhance workforce efficiency, while ensuring efforts are goal-oriented, focusing on specific capabilities to guide processes effectively.

● ● ●

Innovation thrives where purpose leads and barriers fall—
when R&D is given direction without restriction, it becomes
not just a function, but a force for meaningful transformation.

PRINCIPLE 8 – REQUIREMENTS

The word "requirements" often carries an air of formality that can make discussions feel rigid or uncomfortable. Conversations about requirements tend to establish definitive boundaries—delineating what must or must not be done—while leaving little room for flexibility or compromise. When customer or stakeholder requirements are formally documented, the language typically dictates what people must do, how a product should function, and which Key Performance Parameters (KPPs) and Key Performance Indicators (KPIs) must be met. This level of rigor in requirements collection and management is often perceived as bureaucratic, and while that perception holds some truth, well-defined requirements are essential for accurately identifying needs and ensuring effective solutions.

Deliberate Considerations (DC)

Requirements serve as a cornerstone in analyzing technology, processes, and specific desired outcomes. A simple request may highlight a need, but without a well-justified requirement, it lacks the depth necessary to drive meaningful action. Consider an employee approaching a program manager with a request for a new telephone line with additional call features. At face value, this is merely a request. If the justification is that the employee now has an expanded role requiring outbound client calls, this presents a business need—but what is the actual requirement? Through further investigation, the team might determine that the employee needs a multi-line phone with long-distance capability, programmable speed-dial buttons, and a separate voicemail box to effectively perform their role.

Yet, even with this information, the best solution might not be as simple as purchasing a telephone. Additional discussions with a technical team could reveal that the company's legacy phone system is not the only available option. Instead, the organization may have access to an advanced, customizable, computer-based voice communication platform that better integrates with the service ecosystem. Had the decision been made solely on the initial request, the solution might have failed to align with broader business objectives or overlooked existing capabilities. This underscores a critical point—requirements should not just define a solution but also carefully consider the intended outcome and broader organizational impact.

This principle applies across a wide range of industries and operational environments. The focus should not be solely on gathering requirements but also on evaluating their implications and ensuring that the proposed solution aligns with operational priorities. A requirement is more than a checklist of needs—it defines how a solution connects to the business, integrates with existing systems and users, and impacts the broader service environment.

Key Considerations in Requirement Analysis

To ensure a requirement is valid, actionable, and aligned with business needs, organizations must evaluate EPIC FACTs:

- **Environmental Impact** – What are the expected advantages or drawbacks of implementing this requirement?
- **Priority** – Is the requirement critical, or is it a lower-priority enhancement?
- **Intent** – Is this a fundamental need, or is it a preference?
- **Connectivity** – How does this requirement integrate with other business systems, users, or environments?
- **Feasibility** – Is the request practical to accomplish given resource constraints?
- **Accuracy** – Are the specifics clearly defined and measurable?
- **Clarity** – Is the requirement easily understood and implemented by others?
- **Testability** – Can the product or service be evaluated for effectiveness before full deployment?

Deliberate Considerations in Requirements Analysis

Within the Shift framework, requirements are not merely documentation exercises, they drive innovation, strategic decision-making, and service improvements. The following Deliberate Considerations guide the analysis and application of requirements:

- **Innovation** – Does the requirement introduce new ideas, methods, or capabilities that enhance operations?
- **Analysis** – Has the requirement been thoroughly examined to ensure its viability?
- **Integration** – Can the solution be seamlessly incorporated into existing business functions?
- **Simplicity** – Is the requirement structured to be clear, efficient, and easy to implement?
- **Security** – Does the requirement address safety, risk mitigation, and protection of assets?

By adopting a structured yet adaptable approach to requirements gathering, organizations ensure that solutions are not only functional but also strategically aligned with business objectives. The key is not to view requirements as static constraints but rather as dynamic guidelines that facilitate effective decision-making and continuous service improvement.

●●●

DC 31 - Innovation

Innovation is the deliberate process of generating, refining, and implementing new ideas, concepts, or technologies in ways that enhance existing systems, environments, or operational frameworks. While creativity fuels the initial development of innovative solutions, true innovation extends beyond mere ideation—it requires a structured approach to selecting, integrating, and adapting these advancements in a way that drives meaningful improvements.

The ability to incorporate emerging technologies, such as Artificial Intelligence (AI), cloud computing, and smart devices, does not inherently constitute innovation. The real measure of innovation lies in how these technologies are strategically adapted to augment and optimize existing processes. Shift methodology emphasizes this distinction by continuously exploring ways to leverage technology as a force multiplier, ensuring that service delivery is not only enhanced but also remains agile, scalable, and responsive to evolving needs.

While understanding the capabilities and value propositions of advanced technologies is foundational, the real transformative potential is realized through their unique application within a specific service environment. Successful integration is not simply about adopting the latest advancements, it involves aligning these innovations with organizational objectives, operational efficiencies, and customer expectations. This requires an iterative approach, where new capabilities are not only introduced but are also continuously refined to maximize effectiveness and long-term sustainability.

Innovation plays a particularly critical role during the requirements-building phase, where emphasis is placed on inventorying current capabilities and aligning them with needs assessments. By systematically identifying essential tasks and performance outcomes, organizations can effectively reverse-engineer existing capabilities into viable solutions. This structured approach enables Service Managers to reevaluate available technologies and resources, ensuring optimization to meet both present and future operational demands.

Through this lens, innovation becomes a strategic enabler rather than a standalone pursuit—one that integrates cutting-edge advancements while maintaining a clear focus on service effectiveness, operational resilience, and continual improvement.

Worksheet: Innovation and the Requirements Building Phase

Objective: This worksheet is designed to guide Service Managers and teams in identifying and leveraging existing capabilities during the requirements-building phase. By inventorying current resources and aligning them with new needs assessments, teams can reverse-engineer solutions to meet objectives. In the below example, a fictional Corporate Resource Management (CRM) system was created to demonstrate steps.

Step 1: Inventory Current Capabilities

List current capabilities, resources, and tools available within your team or organization. These might include software, systems, personnel skills, methodologies, or workflows.

CAPABILITY	DESCRIPTION	CURRENT USE/ FUNCTION	LIMITATIONS
Example CRM System	Tracks customer interactions and data	Sales and customer support tracking	Limited to customer-facing roles

Step 2: Needs Assessment

Identify and list the key needs or problems that must be addressed within your project or service. Be specific about the performance outcomes required.

NEED/PROBLEM	DESIRED OUTCOME	CURRENT GAP
Example: improve client data sharing	Streamlined real-time access to data	Current system has limited sharing capabilities

Step 3: Match Capabilities to Needs

Now, compare your inventory of current capabilities to the needs you have identified. Explore how these capabilities can be adapted, re-engineered, or combined to form solutions.

NEED/PROBLEM	CAPABILITY MATCH	ADAPTATION/ RE-ENGINEERING OPPORTUNITY
Example: improve client data sharing	CRM System	Re-engineer CRM system for data sharing among internal teams

Step 4: Identify Essential Tasks and Performance Outcomes

Break down the essential tasks that will guide the innovation process and establish performance outcomes. List specific actions that need to occur for the current capability to meet the new requirements. This data can also be utilized to complete a Work Breakdown Structure (WBS).

TASK	PERFORMANCE OUTCOME	WHO IS RESPONSIBLE?
Example: integrate data-sharing tools	Real-time, accurate data sharing between teams	IT Team

Step 5: Use Requirements as a Driver

Once clearly defined requirements are documented, use them as a driver for re-evaluating existing capabilities. Think of the requirements as the "propellant" that pushes the adaptation process forward.

NEW REQUIREMENT	PROPELLANT ACTION	RESULT
Example: Data sharing amount departments	Re-evaluate CRM and communications tools	Development of a custom internal sharing feature

Step 6: Evaluate and Adjust

After implementing solutions, evaluate their effectiveness. Determine if further adjustments are needed to align the adapted capabilities with the desired outcomes.

SOLUTION	EVALUATION	ADJUSTMENT NEEDED? (YES/NO?)
Example: Data sharing feature	Feedback shows improved collaboration	No

This worksheet provides a step-by-step process that can be customized to specific project or program needs, fostering a proactive and solution-driven mindset in Service Management.

In conclusion, innovation within Service Management doesn't always require starting from scratch—it often begins by recognizing and reimagining what's already available. By systematically inventorying existing capabilities and aligning them with clearly defined requirements, teams can reverse-engineer effective, purpose-driven solutions. This structured, six-step approach fosters a proactive mindset, enabling Service Managers to adapt quickly, solve problems creatively, and drive meaningful outcomes through continuous evaluation and refinement.

DC 32 - Integration

To successfully plan for and implement requirements integration while minimizing disruption within existing ecosystems, it is essential to identify, define, and track the core characteristics of each requirement. This structured approach ensures that integration is deliberate, thorough, and aligned with both organizational goals and operational realities. A well-developed Integration Plan checklist serves as a guiding framework, detailing how to introduce solutions into test or operational environments. During this process, it is not uncommon for additional requirements to emerge—needs that neither the customer nor the service team initially anticipated. Recognizing and incorporating these new requirements ensures a seamless and adaptable integration.

PEAK Framework for Requirements Integration

Requirements integration can be effectively assessed using the PEAK framework, which focuses on four critical elements: Practicality, Environment, Applicability, and Knowledge. These elements ensure that integrations are functional, strategic, and sustainable within the service ecosystem.

- **Practicality** – Evaluates whether a pragmatic approach has been taken in integrating the requirement into the ecosystem. The goal is to ensure that the solution meets the identified need without introducing unnecessary complexity or "gold plating," which can obscure the core purpose of the integration. A practical approach ensures that resources are efficiently allocated, and that the final solution remains functional without being over-engineered.
- **Environment** – This element examines how well the proposed solution fits within the existing ecosystem. It is crucial to determine whether the integration is a natural extension of the current system or if it feels forced, potentially leading to disruptions or unintended consequences. A well-aligned solution enhances functionality and ensures long-term sustainability, while a poorly integrated one may introduce inefficiencies or conflicts with existing workflows.
- **Applicability** – Ensures that every component of the Integration Plan directly supports the requirement and its intended outcome. Any items with unclear relevance should be reviewed to confirm their importance and impact. This step keeps the integration process focused, eliminating redundant or unnecessary efforts that might complicate the overall implementation.

- **Knowledge** – This element assesses the depth of understanding applied to the requirement. It includes evaluating the expertise of team members and stakeholders, gathering relevant historical data, reviewing previous initiatives, and leveraging lessons learned from similar integrations. A comprehensive knowledge base minimizes oversight, improves decision-making, and increases the likelihood of successful implementation.

Ensuring a Seamless Integration Process

By addressing these PEAK elements, organizations can create an Integration Plan that is not only structured and well-executed but also accounts for the complexities of Service Management. This approach minimizes risks, aligns solutions with organizational objectives, and ensures that requirements integration is both efficient and adaptable. By incorporating continuous evaluation and stakeholder collaboration, organizations can support long-term success and scalability within their ecosystems.

The Integration Plan checklist is a structured framework designed to guide the integration of requirements into existing ecosystems with minimal disruption and maximum alignment to organizational goals. PEAK represents the key elements to consider when planning and implementing solutions. This approach ensures that integrations are thoughtful, efficient, and sustainable by evaluating the relevance, fit, and impact of proposed solutions. The PEAK checklist helps organizations maintain focus on core objectives while addressing potential challenges and leveraging available resources and expertise, fostering successful and seamless integration efforts.

Integration Plan Checklist

This checklist assists with planning and performing requirements integration to limit disruption within existing ecosystems and ensure smooth transitions of new solutions.

1. Practicality

- Is there a pragmatic approach to integrating the requirement into the ecosystem?
- Does the proposed solution meet the core need without unnecessary complexity or "gold plating"?
- Does the integration avoid overcomplicating or misrepresenting the need?

2. Environment Fit

- Does the integration align naturally with the existing environment?
- Does the solution fit without feeling forced or creating questionable outcomes?
- Are any negative impacts or challenges identified in the environmental relationships?

3. **Applicability**

- Do all components of the Integration Plan relate directly to the proposed solution and requirements?
- Have items with uncertain relevance been reviewed and evaluated for their applicability and impact?

4. **Knowledge Evaluation**

- Have relevant information and information sources been evaluated?
- Is team member experience and expertise considered in the integration plan?
- Have stakeholder inputs been collected, assessed, and integrated?
- Has existing documentation been reviewed thoroughly?
- Have previous internal and external initiatives been researched for lessons learned?

5. **Additional Considerations**

- Have any new requirements arisen during planning that should be addressed?
- Are all stakeholders aware of potential changes to requirements?

Note: Ensure this checklist is revisited and updated during various stages of integration to capture any emerging needs or adjustments to the plan.

In conclusion, successful integration is not simply about combining components—it's about ensuring each element contributes meaningfully and seamlessly to the overall solution. By applying a thoughtful, criteria-driven approach that evaluates fit, applicability, knowledge, and emerging considerations, Service Managers can minimize risk and enhance alignment with both strategic goals and operational realities. Revisiting and refining the integration plan throughout its lifecycle ensures continued relevance, stakeholder engagement, and long-term success.

●●●

DC 33 - Simplicity

When establishing requirements, embracing the synergy of simplicity ensures that solutions remain functional, effective, and manageable. Systems and processes tend to operate more efficiently when they are not overengineered. Reducing complexity, particularly in technological and service-based requirements, leads to solutions that are easier to measure and evaluate against defined success criteria. Simplifying processes improves clarity, reduces confusion, and promotes more efficient outcomes. However, not all scenarios lend themselves to simplification. Some requirements are inherently complex due to factors such as:

- Engineering and design constraints
- Specific regulatory guidelines
- Environmental factors and tolerances

These elements often introduce necessary layers of complexity that must be addressed in documentation. While detailed requirements are sometimes unavoidable, the challenge lies in communicating intricate technological solutions in a way that is accessible and easily understood by all stakeholders. Breaking down complex ideas into simpler, more digestible concepts enhances clarity, ensuring that requirements are not only well-defined but also effectively conveyed.

Understanding the audience is a critical aspect of this process. Whether communicating with stakeholders, customers, or internal teams, Service Managers must tailor their messaging accordingly. Documentation, presentations, and service explanations should be structured in a way that balances technical accuracy with accessibility. The ability to simplify without losing critical details is a valuable skill, as oversimplification can lead to misunderstandings or omissions, while excessive complexity can result in confusion and inefficiency.

A well-rounded Service Manager must refine their ability to communicate at diverse levels, providing detailed explanations where necessary while summarizing when appropriate. Striking the right balance ensures that stakeholders receive the information they need without being overwhelmed or left uninformed. Achieving this equilibrium enables teams to maintain alignment, drive successful outcomes, and foster collaboration across technical and non-technical groups alike. The goal is to ensure that all parties, regardless of their technical background, fully comprehend the requirements and their implications, leading to more seamless execution and effective decision-making.

CASE STUDY: Simplifying Complex Technological Requirements for a New Energy Management System

Background: A large utility company, PowerStream Inc., embarked on a project to develop and implement a new Energy Management System (EMS). The goal was to enhance energy efficiency across multiple facilities by integrating advanced sensors, real-time data analytics, and automated control systems. The project was critical due to regulatory compliance requirements, environmental factors, and the need for precise engineering and design.

Challenge: Requirements for the EMS were inherently complex due to:
Engineering and Design: The system needed to integrate with existing infrastructure, support several types of sensors, and handle high volumes of data with minimal latency.

Specific Regulatory Guidelines: Compliance with industry standards for energy efficiency and data privacy required intricate documentation and validation processes.

Environmental Factors and Tolerances: The system had to operate reliably in varying environmental conditions, including extreme temperatures and humidity levels.

Despite the necessity for detailed requirements, the project team faced challenges in communicating these complexities effectively to key stakeholders, including the executive team, project managers, and technical staff.

Approach:

1. Synergy of Simplicity: To address the challenge, the project team focused on reducing unnecessary complexity in their requirements documentation and communication. They broke down the requirements into core components, highlighting essential functionality and regulatory needs without overwhelming stakeholders with excessive technical jargon.

2. Audience-Centric Communication: The team tailored their communication strategy based on the audience:

- For executives, they provided high-level summaries emphasizing strategic benefits, regulatory compliance, and expected ROI.
- For project managers and technical staff, they presented detailed specifications, implementation plans, and troubleshooting guidelines in a more structured and digestible format.

3. Balancing Detail and Simplicity: The team implemented a layered approach to documentation. Key components and their interactions were described in simplified diagrams and flowcharts, while detailed technical appendices addressed more intricate aspects of the system. This approach helped in maintaining clarity without sacrificing necessary details.

4. Iterative Feedback and Refinement: Regular feedback sessions were held with stakeholders to ensure that the simplified communication was effective and that all parties were aligned with the project goals. Adjustments were made based on feedback to better address concerns and clarify any misunderstandings.

Results:

- Improved Understanding: Simplified communication and documentation facilitated better understanding among stakeholders, effectively reducing confusion while aligning expectations.
- Efficient Implementation: By focusing on core components and essential details, the implementation process became more efficient, and the project team was able to address issues more effectively.
- Successful Compliance: The EMS met regulatory requirements and operated reliably across varying environmental conditions, achieving the desired energy efficiency improvements.

In conclusion, the hypothetical PowerStream Inc. case study underscores that simplicity is not about minimizing detail, but about optimizing clarity for every audience involved. By intentionally integrating layered documentation, audience-specific messaging, and iterative stakeholder feedback, the project team transformed a highly complex initiative into a successfully implemented, regulation-compliant solution. This approach demonstrates that when simplicity is treated as a deliberate strategy—not an afterthought—organizations can enhance understanding, streamline execution, and achieve impactful outcomes without sacrificing technical integrity.

PRINCIPLE 9 – OUTCOME

Measuring the effectiveness of technological applications requires a structured evaluation of how well a given solution aligns with its intended purpose and fulfills the requirements it was designed to meet. Success is determined not only by whether the technology performs as expected but also by its broader impact on service efficiency, usability, and overall organizational goals. Since no implementation is static, outcomes must be continuously refined through consistent measurement, analysis, and adaptation. However, quantifying results is not always straightforward, and Service Managers, along with their teams, may need to adopt creative approaches to establish reliable methods for tracking performance, comparing data, and driving continuous improvement.

Deliberate Considerations (DC)

The Shift methodology approaches technology through a holistic lens, incorporating both People and Process into the overall evaluation of outcomes. Technology, at any level of implementation, generates results that must be measured against predefined requirements to determine its value-add, effectiveness, and overall impact on the service environment. This evaluation extends beyond confirming functionality—it involves identifying areas within the organization that are directly and indirectly affected by the solution. Measuring these outcomes requires an openness to discovering unexpected or undesirable effects, as not all implementations will produce immediately positive results. Recognizing and analyzing negative outcomes should not be seen as a failure but as an opportunity for growth. Every misstep provides valuable data points that can be leveraged to refine processes, improve future applications, and enhance decision-making.

One of the most critical aspects of this evaluation process is fostering a culture of transparency and learning rather than one of fear or blame. Service Managers must resist the tendency to attach punitive measures to undesirable results, instead focusing on constructive analysis and forward-thinking solutions. The Service Cycle should always be considered when designing countermeasures, ensuring that negative outcomes are met with practical refinement rather than reactionary consequences. In most cases, unfavorable results are not the result of intentional negligence or malice but rather unforeseen challenges, gaps in planning, or external variables. If employees fear repercussions for reporting failures, the organization risks creating a culture in which issues are hidden, misrepresented, or ignored. These tendencies obstruct progress and inhibit the ability to adapt and evolve.

By embracing setbacks as learning opportunities, organizations cultivate an environment where technology is continuously optimized, and where teams feel empowered to provide honest assessments without fear of retribution. Measuring the effectiveness of technological applications should be seen as an iterative process, where insights—both positive and negative—serve as the foundation for ongoing enhancement. Through this approach, Service Managers ensure that technology not only meets current needs but also evolves in alignment with future challenges and opportunities.

• • •

DC 34 - Solutions/Results

Results should be carefully observed, systematically measured, and accurately recorded to determine whether a solution has performed as intended according to its defined requirements. A deliberate analysis of both expected and unexpected outcomes can reveal unintended effects that may impact various areas of the organization. The introduction of emergent technologies into an environment has, in many cases, unknowingly led to issues such as privacy violations, disruptions to existing workflows, job displacement, and dependencies on technology that counteract ecosystem stability. Identifying these unintended consequences early is critical to mitigating risks and ensuring that technological integrations support, rather than undermine, operational objectives.

In many cases, the introduction of technology aims to streamline operations, whether by consolidating data, improving visibility into data sources, or automating system access. However, even these functions come with inherent risks that must be continuously evaluated. At the same time, technology may also yield positive unintended side effects, such as process efficiencies, enhanced collaboration, or newly discovered capabilities. These beneficial outcomes should be categorized and explored further to ensure they are effectively incorporated into existing processes, maximizing the full potential of the technology—whether the capability was planned or not.

Documenting Intended vs. Unintended Outcomes

One best practice for tracking and analyzing intended and unintended outcomes is to maintain a Lessons Learned Register. This tool serves as an organized repository for documenting key insights about the performance of a solution, allowing teams to compare projected vs. actual results and assess their broader impact. Capturing these lessons in a structured manner enables organizations to refine their processes, mitigate risks, and build a foundation for future success.

A Lessons Learned Repository should include the following key components:

- Project/Program/Service Information – Identifies the initiative or process where the lesson was observed.
- Observations – Describes key takeaways from the solution's implementation, both positive and negative.

- Recommendations – Provides suggested improvements or actions based on findings.
- Team Member(s) Responsible – Assigns ownership for addressing issues or leveraging insights.
- Actions Taken/Due – Documents corrective or enhancement measures, including their timelines.
- Logging Information – Captures the date of entry and relevant contextual details.

CASE STUDY: Customer Service as a Priority – The Doorman Fallacy

The Doorman Fallacy highlights the need to continuously monitor functions that surround technological implementation. The impact of automated doors was not evident until other areas began to fail. This fallacy demonstrates where the function of a job is narrowed down to one set of tasks and optimized for those alone.

Door attendants have been operating at upscale hotels since before the mid-1800s. While the historical origins are unclear, they have been stylized differently from bellhops and other front lobby staff because their role is distinct and multifaceted. These individuals represented the first impression for customers patronizing hotels, setting the tone for the entire guest experience. With the advent of automatic doors, although these offered elegance and ease of entry, the personal touch of the experience was lost.

Automatic doors began impacting the customer in unintended ways. Although they saved money on labor costs, hotels saw declines in overall customer satisfaction, which affected revenue. This was because door attendants provided more than the simple act of opening and closing doors. Beyond greeting guests upon arrival and keeping the front entrance clean, door attendants also assisted guests with hailing taxis or rideshares and helped carry small luggage items to or from vehicles, tasks that automated doors could not replicate. Additionally, this service offered a layer of security by monitoring the entrance, deterring potential loiterers, and providing reassurance to guests. These human elements, integral to customer experience, were overlooked when the focus shifted solely to automation, underscoring the limitations of technology when critical human functions are disregarded.

Ensuring Process Improvement Through Lessons Learned

Beyond simple documentation, a Lessons Learned Register should serve as a gateway to continuous process improvement. As with all areas within the Shift framework, solutions and outcomes should be subject to ongoing scrutiny and refinement. The data collected from past implementations should be regularly reviewed to ensure that insights contribute to long-term improvements rather than remaining isolated reflections. Integrating the process improvement cycle into every aspect of Service Management allows for consistent refinement and strategic evolution.

Each entry within the register should contain structured elements that ensure clarity, traceability, and actionability. Unique identifiers facilitate easy tracking, while logging dates provide a timeline for insights and corrective actions. Categorizations, such as Process, Technology, Communication, Resource, or Training, help organize findings and identify patterns across different domains. A root cause analysis ensures that teams not only document issues but also understand the factors contributing to them, allowing for precise solutions.

Furthermore, assigning action owners reinforces accountability, ensuring that recommendations are not only identified but actively pursued. A status tracker, with labels such as Not Started, In Progress, or Completed, provides visibility into implementation efforts and helps maintain momentum.

● ● ●

A well-maintained Lessons Learned Register transforms past experiences into future successes. By capturing insights systematically and integrating them into decision-making processes, organizations can prevent repeated mistakes, refine strategies, and drive continuous improvement.

By institutionalizing this structured approach, organizations foster a culture of learning and adaptability where lessons from previous initiatives directly contribute to future efficiencies and innovations. This ensures that every technology integration—whether successful or requiring improvement—becomes an opportunity for growth, refinement, and sustained operational excellence.

●●●

DC 35 - Value Add (Money, People, Resources)

Measuring outcomes when introducing new technology into an environment requires a structured evaluation of value determinations—assessing how the output or new state compares to the previous status quo. Value-add is the additional advantage created through this transition, representing an improvement in efficiency, capability, or overall effectiveness. The most reliable method for measuring value-add is through established metrics and business process evaluations conducted before, during, and after the integration of new technology. These measurements provide quantifiable insights into how well the new system performs in comparison to prior methods.

A thorough examination of current-state metrics creates a historical data repository over time. This data serves as the foundation for trend analysis, allowing organizations to measure present performance, compare it against past benchmarks, and develop projections for future outcomes. Understanding the trajectory of technological advancements within a service environment is key to strategic planning and long-term growth. With a well-documented historical record, decision-makers can establish clear pathways for enhancing capabilities and determine where the program is heading in the coming years. Value-add, in this context, refers to the measurable impact of innovative technologies that contribute to an elevated level of performance or an upward technology curve.

Defining metrics and measurable attributes is essential for evaluating the effectiveness of technology implementations and calculating the value added. These measures should be designed to provide clear visibility into whether a new solution has enhanced business operations, improved efficiency, or optimized resource development. Value is often assessed through resource improvement, which can encompass assets, facilities, personnel, and operational structures. Resources may be tangible, such as time, revenue, and physical assets, or intangible, such as brand reputation, intellectual capital, and industry influence.

When direct metrics are not available, facilitating discussions and posing key questions can help define and quantify value-add. These inquiries should focus on identifying measurable benefits, mitigating unintended consequences, and ensuring that any improvements do not create new inefficiencies elsewhere. Some key considerations include:

- Has this implementation increased revenue, saved time, or reduced overhead in producing a product or report?

- Did the improvements in one area result in a loss or inefficiency in another?
- Could changes in processes negatively impact another team that relies on the current output, leading to an overall decrease in value?
- Are there any direct or indirect effects that could influence the broader service ecosystem, either positively or negatively?

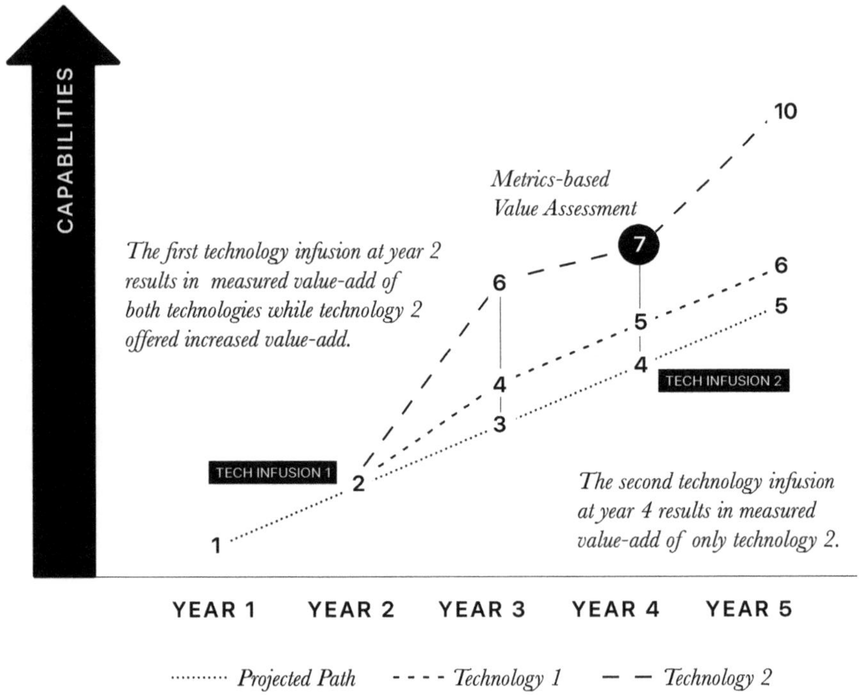

By consistently measuring and analyzing these factors, organizations can ensure that recent technological changes drive meaningful progress without introducing unintended drawbacks. Understanding how value is added across various aspects of the service environment enables more informed decision-making and fosters a culture of continuous improvement.

What the Chart Communicates

1. Incremental vs. Step-Change Improvements

- Each "infusion" triggers a step-like increase in capability rather than a smooth, gradual climb. This highlights that new technology investments often result in immediate performance jumps, though the size and duration of the improvement can vary.

2. Comparing Multiple Technologies

- By plotting Technology 1 and Technology 2, the image underscores that not all solutions adopt equally well over time. One technology may stagnate while another continues to grow (or vice versa), and some solutions respond more strongly to certain investments than others.

3. Continual Monitoring

- The "Metrics-Based Value Assessment" at around Year 3 suggests a deliberate checkpoint to measure how these technologies perform in real-life operations. This signals the importance of periodically evaluating whether a new tool or process is delivering on its promise.

4. Long-Term vs. Short-Term Effects

- Over the five-year span, you see both short-term gains (right after the infusion) and longer-term trends (whether the technology continues to climb, plateaus, or responds to subsequent infusions).
- The baseline "Projected Path" helps visualize the added benefit: how much better things are compared to simply "doing nothing new."

5. Value-Add Fluctuations

- Not every infusion benefits every technology equally. Technology 2, for example, apparently benefits more from the second infusion (Year 4) with the value of 7, while Technology 1's curve remains relatively flat. This underscores how a given strategy or investment may not universally upgrade all solutions in play.

Why It Matters in a Service-Improvement or Technology-Adoption Context

- Decision Points and ROI - This chart framework is often used by Service Managers or leadership to decide when and why further investments are justified—and to measure whether each investment delivers the intended value.
- Different Adoption Curves - Technology solutions can diverge significantly in how quickly or effectively they scale up. The image teaches that you cannot treat all solutions the same; each path is unique and may require different degrees of support, training, or follow-up investment.
- Importance of Midstream Assessments - Scheduled value assessments (like the one shown at the top during Year 3 and Year 4) help catch issues early and provide a chance to pivot or double down on promising areas. It is a reminder of continuous monitoring—not just a post-project "lessons learned."

- Realistic Expectations for Stakeholders - The image clearly shows that immediate leaps in capability can be followed by slower climbs or plateaus. This realism is crucial in managing stakeholder expectations for how quickly an organization can realize benefits from new tools or processes.

In conclusion, delivering true value in service environments requires more than just implementing new technologies or processes—it demands a thoughtful assessment of how those changes impact the broader ecosystem of people, resources, and outcomes. By continuously monitoring value-add across multiple dimensions and recognizing that gains in one area can inadvertently introduce friction elsewhere, organizations can make more informed, balanced decisions. The emphasis on midstream evaluations, adoption variability, and realistic performance expectations ensures that investments yield sustained benefits, align with stakeholder priorities, and reinforce a culture of continuous, measurable improvement.

● ● ●

DC 36 - Requirements Fulfilled

Revisiting requirements is essential to determine if a technology aligns with its overall purpose and intended application by comparing outcomes to predefined expectations. Despite a project's best efforts, developments often deviate from their initial design, leading to what is commonly known as scope creep. While scope creep is typically avoided whenever detected, it can sometimes go unnoticed until it is too late. However, not all instances of scope creep result in negative outcomes. In fact, these deviations can often create unexpected opportunities. It is crucial to recognize that unplanned benefits and unintended capabilities may emerge once a new technology is fully operational within its environment.

Rather than hastily eliminating these positive outcomes, Service Managers should conduct a deliberate assessment of the technology's capabilities against the original requirements. This evaluation enables them to map and correlate new functionalities to program initiatives and existing processes within the ecosystem. Once this comprehensive assessment is completed and documented, the process of continuous evaluation begins. The ongoing assessment of requirements, which now serve as performance parameters, is key to ensuring the technology continues to meet the mission's evolving needs.

Continuous assessment is more than just a one-time evaluation; it is a proactive loop where technology is consistently measured against both current and emerging requirements. Through this ongoing analysis, Service Managers may uncover unexpected impacts or potential failures that might not become evident immediately after implementation. Only through time and repeated assessment can the full scope of technology's effect on other areas be understood. This iterative process ensures that technology remains agile and adaptable, capable of addressing both intended and unforeseen challenges as they arise within a dynamic service environment.

Moreover, this ongoing cycle of evaluation and adaptation fosters a culture of Continuous Improvement. As Service Managers gather insights from their assessments, they can identify trends and patterns that inform future decisions and strategies. By staying attuned to both the successes and shortcomings of the technology, managers can make data-driven adjustments that enhance overall service delivery. This proactive stance not only improves individual technology solutions but also strengthens the service ecosystem, promoting innovation and efficiency across the board. In this way, technology not only fulfills its original intent but also evolves to meet the changing demands of the organization and its stakeholders. Ultimately, embracing this comprehensive approach to continuous assessment transforms potential setbacks into valuable lessons, driving the organization toward greater effectiveness and resilience in a fast-paced and ever-evolving landscape.

Revisiting and continuously assessing technology against original requirements is essential to ensure alignment with mission goals and to identify both scope creep and unplanned benefits. While deviations can present risks, they may also reveal new opportunities. Through iterative evaluation, Service ManagerService Managers can adapt technology to evolving needs, uncover hidden impacts, and drive continuous improvement. This proactive, data-driven approach enhances service delivery, fosters innovation, and ensures long-term relevance in dynamic environments.

Exercise: Outcome Evaluation and Process Adjustment

Objective:
This exercise helps readers reflect on past project outcomes to identify both intended and unintended consequences. By analyzing these outcomes, participants will propose adjustments to refine future processes and achieve better results.

Part 1: Select a Past Project

1. Choose a Project: Reflect on a completed project from your experience. The project should have a clear objective and measurable results.
Example: Implementing a new customer service system, launching a marketing campaign, or introducing a new product line.
2. Document the Original Goals: Write down the project's intended goals and expected outcomes.
Example: "Improve customer response times by 20% within three months."

Part 2: Analyze the Outcome

1. Assess the Results: Compare the intended goals to the actual results. What was achieved? What fell short?
Example: "Response times improved by 15%, but customer satisfaction ratings remained static."
2. Identify Unintended Consequences: Reflect on any outcomes not originally anticipated—positive or negative.
Example Unintended Consequences:
Positive: Employees developed new skills while adapting to the new system.
Negative: Increased workload led to staff burnout.
3. Evaluate Contributing Factors: Analyze the factors that led to both the intended and unintended consequences.
Example: "The training period was shorter than needed, leading to initial confusion among staff."

Part 3: Propose Adjustments

1. Address Unintended Consequences: For each unintended consequence, propose specific adjustments to processes or strategies.

Example Adjustments:

- *Introduce phased training sessions to avoid overwhelming staff.*
- *Implement a feedback loop during the rollout phase to identify pain points early.*

2. Refine Future Processes: Suggest broader improvements to the overall process *based on lessons learned.*

Example: "Incorporate stakeholder input earlier to identify potential challenges before implementation."

Part 4: Reflection and Discussion

1. Reflect on Insights: Answer the following questions:

- What could have been done differently to prevent negative consequences?
- How can the adjustments you proposed enhance future project outcomes?

2. Group Discussion (Optional): If working with a team, share your analysis and proposed adjustments. Gather additional insights from peers to refine your approach further.

By the end of this exercise, readers should produce:

- A clear comparison of the project's intended and actual outcomes.
- A list of unintended consequences and their root causes.
- Proposed adjustments to improve processes for similar future projects.

This activity equips readers with a structured approach to evaluating outcomes and making thoughtful adjustments to ensure continuous improvement and better alignment with organizational goals.

● ● ●

Shift Roadmap:
A Visual Guide to Service Improvement

---

1. Identify the Need for Change

- Catalyst Recognition: Identify triggers prompting the need for improvement (e.g., customer feedback, inefficiencies, technological advancements).
- Define Goals: Establish clear objectives for the change, aligned with organizational and customer priorities.

2. Evaluate the Current Environment

- Assess the Service Ecosystem: Understand the interplay of people, processes, and technology in the current environment.
- Gap Analysis: Pinpoint deficiencies, redundancies, or opportunities for enhancement.
- Stakeholder Engagement: Collaborate with key stakeholders to gather insights and build alignment.

3. Plan the Improvement

- Develop a Strategy: Create an actionable plan that aligns with Shift principles and considers Deliberate Considerations (DCs) such as simplicity, practicality, and adaptability.
- Set Measurable Targets: Define Key Performance Indicators (KPIs) and success metrics to evaluate progress.
- Resource Allocation: Determine the resources (time, budget, personnel, technology) required for implementation.

4. Execute with Purpose

- Implement Changes: Introduce adjustments to people, processes, or technology incrementally.

- Foster Collaboration: Ensure open communication among team members and stakeholders to support smooth implementation.
- Iterate as Needed: Use feedback loops to refine and improve the process during execution.

5. Monitor and Measure Progress

- Track KPIs: Continuously measure performance against predefined targets.
- Collect Feedback: Gather input from stakeholders, customers, and team members to evaluate the effectiveness of changes.
- Identify Outcomes: Document both intended and unintended results for analysis.

6. Optimize and Adapt

- Analyze Results: Assess the impact of changes on the service ecosystem and identify areas for further improvement.
- Refine Processes: Apply lessons learned to enhance workflows, align outcomes with goals, and ensure sustainability.
- Scale Solutions: Adapt successful changes for broader application across the organization.

7. Foster Continuous Improvement

- Embed a Learning Culture: Encourage ongoing evaluation, innovation, and refinement.
- Leverage Technology: Integrate emerging technologies to drive efficiency and scalability.
- Update the Service Wheel: Revisit and iterate on the improvement cycle to maintain relevance and effectiveness.

Shift Roadmap at a Glance

1. Catalyst Recognition – Identify the need for change.
2. Environment Evaluation – Analyze the current service ecosystem.
3. Strategic Planning – Develop a clear, actionable improvement strategy.
4. Purposeful Execution – Implement changes with collaboration and iteration.
5. Performance Monitoring – Measure progress and gather feedback.
6. Process Optimization – Refine changes for better outcomes and scalability.
7. Continuous Improvement – Foster a culture of learning and adaptation.

This roadmap ensures a clear and structured journey for practitioners to navigate service improvement using the Shift methodology.

CONCLUSION

Shift is more than just a methodology; it is a philosophy of continuous improvement designed to empower organizations to refine their approach to Service Management by making deliberate, impactful adjustments to the interplay of people, processes, and technology. Rooted in actionable principles, Shift provides a structured framework to address challenges, foster innovation, and promote adaptability across service ecosystems. At its core, Shift recognizes that service excellence is driven by the synergy of its foundational elements: people, whose diverse talents and perspectives create the foundation for innovation; processes, which provide the structure needed for consistent and scalable outcomes; and technology, the enabler that amplifies capabilities and introduces new efficiencies. Through Deliberate Consideration of these elements, Shift challenges organizations to think beyond conventional solutions, inspiring practitioners to embrace adaptability, encourage collaboration, and prioritize outcomes that drive both organizational success and customer satisfaction.

Shift is not merely about incremental improvements; it aspires to create an ecosystem of perpetual growth and learning. By focusing on key principles such as alignment, communication, engagement, adaptability, catalysts for change, measurable targets, and purpose-driven innovation, organizations can sustain progress even in the face of rapid change. It serves as a compass for navigating complexity, ensuring that every action contributes to a well-rounded and resilient service environment. It equips teams to identify and address systemic gaps, enhance service quality through iterative adjustments, foster meaningful connections between stakeholders and processes, and leverage technology thoughtfully to drive efficiency and innovation.

Importantly, Shift is not static; it evolves with its practitioners. As new challenges emerge and industries advance, its principles remain a guide for uncovering opportunities and achieving excellence. Practitioners are encouraged to apply the lessons of Shift creatively, adapting the methodology to suit unique organizational contexts and goals. Emphasizing the importance of reflection and feedback, Shift thrives on the commitment to question, refine, and innovate. The continuous improvement cycle ensures that successes are built upon, failures are addressed constructively, and every experience contributes to a growing repository of insights.

As you step forward, let Shift inspire you to lead with purpose, adaptability, and a mindset of growth. Embrace its tenets to transform challenges into opportunities and create a legacy of excellence in service delivery. Remember, the journey of improvement is never complete. With every small "Shift," you contribute to a larger, more meaningful impact. Use this methodology to ask better questions, seek deeper understanding, and deliver exceptional results. Together, let us shape a future where service excellence is not just an aspiration but a standard.

Glossary of Terms for Shift

Adaptability: The ability to adjust processes, strategies, or systems in response to new challenges, changing environments, or evolving customer needs. A key principle in Shift for ensuring services remain effective and relevant.

Alignment: The process of ensuring that people, processes, and technology are working harmoniously toward shared organizational goals and customer objectives.

Catalyst: The trigger or driving force behind a need for change, which prompts organizations to initiate process improvements or innovations.

Continuous Improvement: An ongoing effort to enhance services, products, or processes through incremental changes, feedback loops, and iterative refinement.

Deliberate Considerations (DC): Specific, actionable insights within the Shift framework that guide decision-making and process improvements. Examples include adaptability, practicality, simplicity, and enrichment.

Engagement: The depth and quality of interactions between team members, stakeholders, and customers. Engagement fosters collaboration, commitment, and meaningful contributions.

Environment: The broader ecosystem in which services are delivered, encompassing organizational culture, customer expectations, and external factors like market conditions or technological advancements.

Holistic Service Management: An integrated approach that views people, processes, and technology as interconnected components driving service excellence and operational efficiency.

Innovation: The creation or application of new ideas, methods, or technologies to improve processes, services, or outcomes. A critical element in adapting to evolving needs.

Iteration: A repetitive cycle of testing, feedback, and improvement that allows for incremental changes and continuous refinement of processes or solutions.

Key Performance Indicator (KPI): Quantifiable measure used to evaluate the success of a process, service, or initiative in meeting defined objectives.

Key Performance Parameter (KPP): Benchmarks intended to determine if a system or process meets operational capabilities. KPPs are often utilized during development, testing, and evaluation.

Lessons Learned Register: A structured document used to capture insights, observations, and recommendations from past projects or initiatives to inform future improvements.

Outcomes: The results or impacts of a process, service, or change initiative. Outcomes can be intended or unintended, and both are evaluated for their influence on the service ecosystem.

Practicality: A focus on realistic, achievable solutions that balance innovation with the constraints and capabilities of the environment.

Processes: The structured steps and workflows designed to achieve consistent and repeatable results within a service environment.

Requirements: The necessary attributes, features, or capabilities of a solution, defined to ensure alignment with organizational goals and customer needs.

Service Ecosystem: The interconnected network of people, processes, and technology that drive an organization's service delivery and operational outcomes.

Service Wheel: A conceptual framework representing the cyclical process of assessing, planning, implementing, and monitoring. The Service Wheel is designed to optimize services through continuous improvement and focuses on the customer experience.

Shift: A service improvement methodology designed to implement strategic, incremental changes in people, processes, and technology to enhance service delivery.

Simplicity: The principle of reducing unnecessary complexity to improve clarity, efficiency, and ease of implementation in processes and solutions.

Stakeholders: Individuals or groups with a personal stake in the outcomes of a service, process, or initiative. This includes employees, customers, partners, and other collaborators.

Targets: Clear, measurable objectives that define the desired outcomes of a process or initiative. Targets serve as benchmarks for evaluating success.

Technology: Tools, systems, and platforms used to enhance service delivery, improve efficiency, and support people and processes.

Transparency: The practice of making processes, data, and decisions visible and accessible to all relevant stakeholders, fostering trust and collaboration.

Value Add: The additional benefit created by a process, technology, or service that exceeds the original baseline or status quo.

This glossary encapsulates key concepts and terminology used within the Shift methodology to ensure clarity and understanding for practitioners.

●●●

www.ingramcontent.com/pod-product-compliance
Lightning Source LLC
LaVergne TN
LVHW050914210926
848482LV00012B/433

* 9 7 9 8 2 1 8 7 9 0 7 7 6 *